测绘专业

词汇手册

English-Chinese / Chinese-English Vocabulary Handbook for Geomatics Engineering

主编 尹 晖

主审 Spiros Pagiatakis

编委 尹 晖 张小红 赵建虎 徐 芳 乔俊军 胡春春

WUHAN UNIVERSITY PRESS

武汉大学出版社

图书在版编目(CIP)数据

英汉汉英测绘专业词汇手册/尹晖主编;Spiros Pagiatakis 主审.—武汉:武汉大学出版社,2008.6
ISBN 978-7-307-06227-6

Ⅰ.英…　Ⅱ.①尹…　②S…　Ⅲ.测绘学—词汇—手册—英、汉　Ⅳ.P2-62

中国版本图书馆 CIP 数据核字(2008)第 063609 号

责任编辑:王金龙　　责任校对:程小宜　　版式设计:马　佳

出版发行:**武汉大学出版社**　(430072　武昌　珞珈山)
(电子邮件:wdp4@whu.edu.cn 网址:www.wdp.com.cn)
印刷:湖北民政印刷厂
开本:787×1092　1/32　印张:18.875　字数:421 千字　插页:1
版次:2008 年 6 月第 1 版　　2008 年 6 月第 1 次印刷
ISBN 978-7-307-06227-6/P·135　　定价:36.00 元

内容提要

本书采用英汉-汉英方式收录了测绘工程专业所涉及的各个领域(包括工程测量、大地测量、摄影测量、遥感、制图学、地理信息系统、城市空间信息工程、海洋测绘、数理统计与测量平差等)的相关词汇和基本术语,约有16000条,词汇具有专业性强,涵盖面广,查阅方便等特点,可供测绘工程及相关专业科技人员和高等学校使用。

前　言

为了紧跟测绘学科的国际发展前沿，满足测绘人才走出国门，参与国际学术交流的需要，我们在《测绘工程专业英语》教材的基础上，组织编写了这本《英汉汉英测绘专业词汇手册》(English-Chinese/Chinese-English Vocabulary Handbook for Geomatics Engineering)，本书具有以下特点：

(1) 内容涵盖面广，涉及了测绘工程专业各个领域（包括工程测量、大地测量、摄影测量、遥感、制图学、地理信息系统、城市空间信息工程、海洋测绘、数理统计与测量平差等）的相关词汇和基本术语，约有16000条。

(2) 收词尽可能新，考虑测绘学科的迅速发展，词汇收录参考了近期出版发表的国外原版教材、论文、期刊、国内同类书籍和网络资源，尽可能做到跟踪前沿，广泛收集，丰富积累，规范表述。

(3) 查阅方便实用，词汇采用英汉-汉英方式，按字母顺序编排，简明实用，开本便于携带。

本书由尹晖担任主编，参加编写的人员有：张小红、赵建虎、徐芳、乔俊军、胡春春，加拿大约克大学Spiros Pagiatakis教授和武汉大学尹晖教授负责审定词汇表述的专业性与规范性，最后由尹晖统稿。在编写过程中，得到测绘学院杨睿、李鹏、李潇和许多老师的热情帮助，在此谨致以衷心的感谢。

限于我们的水平，书中不当之处，恳请读者批评指正。

Preface

This dictionary has been designed to facilitate easy, fast and accurate use of geomatics technical terms between English and Chinese.

Geomatics is a rapidly developing engineering field and new specialized technical terms are being introduced almost on a daily basis. Therefore, we placed more emphasis on the contemporary terms and have tried to make this dictionary valid for both languages, but mostly for Chinese students and scholars, who are involved in international projects, participate in international conferences, publish in scholarly journals, use English literature or study abroad in English speaking countries. The collection of all the technical terms included herein would not have been possible without the hard work of our colleagues from the School of Geodesy and Geomatics at Wuhan University, who collected comprehensive lists from their area of expertise. We thank them very much for their input!

As usual, we believe that there is always room for

improvement and we encourage the users of this dictionary to communicate to us any omissions or errors that might have insidiously escaped our scrutiny.

March 2008

目　录

英 汉

A

abbreviation 简写符号

abrupt slope 陡坡

abscissa 横坐标

abscissa axis 横坐标轴,X 轴

abscissa of image point 像点横坐标

absolute accuracy, absolute precision 绝对精度

absolute altitude 绝对高程

absolute average error 绝对平均误差

absolute coordinate 绝对坐标

absolute error 绝对误差

absolute flying height 绝对航高

absolute gravimeter 绝对重力仪

absolute gravity 绝对重力

absolute gravity measurement 绝对重力测量

absolute maximum 绝对极大值

absolute minimum 绝对极小值

absolute orientation 绝对定向

absolute parallax 绝对视差

absolute perturbation 绝对摄动

absolute position 绝对位置

absolute positioning 绝对定位

absolute salinity 绝对盐度

absolute temperature 绝对温度

absolute threshold	绝对阈
absolute value	绝对值
absolute(point)error ellipses	绝对点位误差椭圆
absorber	吸声器
absorption	吸收
abstract data type	抽象数据类型
abstract symbol	抽象符号
abutment	桥台,桥墩
abyssal hill	深海丘陵
acceleration	加速度
acceleration of gravity	重力加速度
accelerometers	加速计
accepted tolerance	规定限差
accident error	随机(偶然)误差
accidental error	偶然误差
accumulation of error	累积误差
accuracy	准确度
accuracy assessment	精度评估,精度检验
accuracy design	精度设计
accuracy for horizontal control	平面控制精度
accuracy of angle	角度精度
accuracy of angular measurement	测角精度
accuracy of break through	贯通精度
accuracy of drawing	绘图精度
accuracy of instruments	仪器精度
accuracy of network	控制网精度
accuracy of ranging	测距精度
accuracy of sea water salinity	盐度准确度

A

accuracy of sea water temperature	水温准确度
accuracy of vertical control	高程控制精度
accuracy test	精度检验
accuracy testing	准确度测试
accurate measurement	精确量测
accurate rectification	精纠正
achromatic film	盲色片
acoustic beacon	声标
acoustic correlation log	声相关计程仪
acoustic current meter	声学海流计
acoustic Doppler current profiler (ADCP)	声学多普勒海流剖面仪
acoustic holography system	水声全息系统
acoustic hull mounted current meter	走行式声学海流计
acoustic image	声图图像
acoustic positioning	水声定位
acoustic positioning system	水声定位系统
acoustic reflector	声反射器
acoustic responder	水声应答器
acoustic scattering	声散射
acoustic signal	声信号
acoustic sounder	回声测声器,回声探测器
acoustic sounding	回声探测
acoustic transponder	水声应答器
acoustic water level	声学水位计
acoustic window	透声窗

acquisition time	初始定位时间
actinograph	辐射计,光测测定仪,自记曝光计
activation function	激活函数
active contour model	主动轮廓模型
active leg	激活航线
active microwave remote sensing	主动微波遥感
active microwave sensors	主动微波遥感传感器
active remote sensing	主动式遥感
active sensor	主动式传感器
active sonar	主动声呐
active tectonic deformation	活动(地质)构造变形
active vision	主动视觉
actual area	实际面积,有效面积
actual distance	实际距离
actuator	调节器,传动装置,激励器,执行元件
acute angle	锐角
acute angle triangle	锐角三角形
adaptation	自适应
adaptive arithmetic coding	自适应算术编码
adaptive disparity estimation	自适应视差估计
adaptive filter	自适应滤波
adaptive least squares correlation	自适应最小二乘相关
adaptive quantization	自适应量化
adaptive resonance theory(ART)	自适应共振理论
adaptor	转接器,拾音器,接合器

ADCP	声学多普勒海流剖面仪
addition	加法
addition constant	加常数
additional charge	附加费用
additional control point	附加控制点
additional parameter's(APs)	附加参数
additive color viewer	彩色合成仪
additive property	可加性
address coding	地址编码
address geocoding	地址地理编码
address matching	地址匹配
adit	横坑,入口
adit of the tunnel	隧道入口
adit planimetric map	坑道平面图
adit prospecting engineering survey	坑探工程测量
adjacent angle	邻角
adjacent effect	邻近效应
adjacent flight line	相邻航线
adjacent map	接边地图
adjacent side	邻边
adjoint matrix	伴随矩阵
adjust to zero	置零,归零
adjusted angle	平差角
adjusted elevation	平差高程
adjusted quantity	平差量
adjusted value	平差值
adjustment	平差,校正,调整

adjustment by directions	方向平差
adjustment by method of junction point	结点平差
adjustment computation	平差计算
adjustment curve	校正曲线
adjustment method	平差法
adjustment of angles	角度平差
adjustment of astrogeodetic network, astrogeodetic adjustment	天文大地网平差
adjustment of correlated observations	相关平差
adjustment of observations	测量平差
adjustment of position	位置平差
adjustment of traverse	导线平差
adjustment of typical figures	典型图形平差
adjustment with rank deficiency	秩亏平差
administrative map	行政区划图
administrative region of a city	市域
ADS40	Leica 经销的机载数字传感器(三线阵数码相机)
advection fog	平流雾
advection parameter	平流参数
advisory committee	顾问委员会
aerial	航空的,空中的,天线
aerial camera	航空摄影机
aerial camera lens	航摄仪镜头
aerial camera mounting	航空摄影装置

A

aerial coverage	航摄资料地区
aerial digital camera	航空数码相机
aerial equipment	航测设备
aerial film	航摄软片
aerial image	航空影像
aerial leveling	航空水准测量，航空抄平
aerial map	航摄图
aerial mapping photography	航测摄影
aerial photo	航片
aerial photo interpretation	航摄图片判读
aerial photogrammetry	航空摄影测量学
aerial photogrammetry	航空摄影测量
aerial photographic camera	航摄仪，航空摄影机
aerial photographic gap	航摄漏洞
aerial photography	航空摄影
aerial photography norm	航摄规范
aerial photomapping	航空摄影制图
aerial platform	航空平台
aerial remote sensing	航空遥感
aerial route map	航线图
aerial spectrograph	航空摄谱仪
aerial strip survey	航线测量
aerial surface	航摄面
aerial surveying camera	航摄仪，航空摄影机
aerial surveying camera	航测摄影机
aerial topographic map	航测地形图
aerial triangulation	空中三角测量

A

aerial triangulation block 空中三角网模型
aerial-camera-calibrated focal length 航摄仪校准焦距
aerocartograph 航空测图仪
aerolux 光度计
aeronautical 航空的,航空导航的
aeronautical chart 航空导航图
aeronautical charting 航空制图
aerophotogeodesy 航空摄影测量,航空摄影大地测量
aerophotogrametry bathymetric system 航空摄影测深系统
aerophotogrammetric survey 航空摄影测量
aerophotogrammetric(al) 航空摄影测量的,航测的
aerophotogrammetry 航空摄影测量
aerophotograph 航空像片,航摄像片
aero-photograph film 航摄胶片
aerophotographic camera 航空摄影机
aerophotographic negative 航摄负片
aerophotographic plan 航测平面图
aerophotographic reconnaissance 航摄勘测
aerophotographic(al) 航空摄影的,航摄的
aerophotography 航空摄影学,航空摄影
aerophototopography 测图航空摄影
aeroplane location 飞行器定位
aerotriangulation 空中三角测量

A

affine rectification	仿射纠正
affine transformation	仿射变换
aggregate volume	总方量
agro-meteorological model	农业气象模型
AI	人工智能
aid measuring pole	辅助测量杆
aid to navigation	航标
aid to navigation	助航标志
AIMS	美国 Automatric 公司的卫星遥感测图处理系统
air base	摄影基线
air base inclination	空中基线倾角
air baseline	空中基线
air drain	通风孔
air shaft	通风竖井
airborne	空运的,空降的,机载的
airborne gravimetry	航空重力测量
airborne laser mapping	机载激光测图
airborne laser scanning	机载激光扫描
airborne laser sounding	机载激光测深
airborne laser sounding system	机载激光测深系统
airborne laser terrain mapping (ALTM)	机载激光地形测图
airborne magnetic bathymetric system	机载磁测深系统
airborne multispetral remote sensing	机载多光谱遥感测深系统

A

airborne remote sensing	航空遥感
airborne SAR	机载合成孔径雷达
airborne senor	机载遥感器
airborne TLS CCD	机载三线阵 CCD
aircraft camera	航摄仪,航空摄影机
airport recognition	机场识别
air-to-ground positioning	空地定位
alarm device	报警装置
alarm system	报警系统
albedo	反照率
aliasing	混叠
alignment	定线,准直
alignment design	(道路)定线设计
alignment of sounding	测深定线
alignment survey	定线测量
alkaline battery	碱性电池
all weather terrestrial rangefinder	全天候测距仪
allowable error	容许误差
allowable value	容许值
almanac	历书,概略星历
almucantar	平纬圈,等高圈
ALOS	微波遥感卫星(2003年发射)
altas information system	地图集信息系统
alteration	改建
alternate angle	错角
alternate plan	备选方案,(线路)比较方案

A

alternating current	交流电
alternative hypothesis	备选假设
alternative route	备选线路
altimeter	测高仪
altimetric data	高程数据
altimetric measurement	测高
altimetric point	高程点
altimetry	测高法,高程测量法
altitude	高度,地平纬度
altitude angle	高度角
altitude correction	高程改正
ALTM	机载激光地形测图
ambiguity	模糊度
ambiguity resolution	模糊度解算
ambiguity resolution technique	模糊度解算技术
amend	修正
amount of rainfall	降雨量
amplifier	放大器
amplitude	幅,振幅,范围
amplitude dispersion index	振幅离差指数
amplitude of oscillation	摆幅,振幅
amplitude of partial tide	分潮振幅
anaglyphic map	互补色地图
anaglyphic plotter	互补色立体测图仪
anaglyphical stereoscopic viewing	互补色立体观察
anaglyphoscope	互补色镜
analog aerotriangulation	模拟空中三角测量
analog map	模拟地图

analog photogrammetry	模拟摄影测量
analog seismometer	模拟磁带地震仪
analog stereoplotter	模拟立体测图仪
analogue	模拟
analogue photogrammetry	模拟摄影测量
analogue/digital converter	模数转换器
analysis of control networks	控制网分析
analysis of image profile	灰度剖面分析
analysis of satellite resonances	卫星共振分析
analysis of variance	方差分析
analytical aerial triangulation	解析空中三角测量
analytical aerotriangulation	解析空中三角测量
analytical geometry	解析几何
analytical image matching system (AIMS)	美国 Automatric 公司的卫星遥感测图处理系统
analytical map	分析地图
analytical mapping	解析测图
analytical mapping control point	解析图根点
analytical orientation	解析定向
analytical photogrammetry	解析摄影测量
analytical plotter	解析测图仪
analytical plumb-line calibration	解析铅垂线检校法
analytical rectification	解析纠正
anchorage chart	锚地图
anchorage survey	锚地测量
ancient map	古地图
ancillary buliding	附属建筑物

ancillary facility	附属设施
ancillary structure	附属构筑物
anemograph	风速记录仪
anemometer	测风仪,风速风向仪,风速计
aneroid barometer	空盒气压计
angle bisector	分角线
angle calculation	角度计算
angle closing error of traverse	导线角度闭合差
angle method	小角度法
angle of convergence	收敛角
angle of deviation	偏角,磁偏角
angle of drift	航偏角
angle of projection	投射角
angle of rotation	旋转角
angle of the sector	扇形角
angle sum of a triangle	三角形内角和
angle-to-right	右转角
anglular closure of traverse	导线角度闭合差
angular accuracy	测角精度
angular calibration	角度检定
angular closure	角度闭合差
angular distortion	角度畸变
angular error	角度误差
angular field of view	像场角
angular momentum	角动量
angular observation	角度观测
angular or distance intersection	角度距离交会

angular orientation	角定向
angular reduction	角度归算,方向改化
angular resolution	角分辨率
angular second moment(ASM)	角度方向二阶矩
angular velocity	角速度
anisotropic diffusion	各向异性扩散
ANN	人工神经网络
annexed leveling line	附合水准路线
annotation	调绘
annotation	注记
annual magnetic change	周年磁变
annual mean sea level	年平均海面
annual parallax	周年视差
annual precession	周年岁差
annual rainfall	年降雨量
anomalous potential	异常位,扰动位
ant algorithm	蚁群算法
antarctic circle	南极圈
antenna	接收机天线
antenna array	接收机阵列
antenna height	天线高度
antenna phase center	天线相位中心
antenna splitter	信号分频器
anthropogenic causes	人为原因
anthropogenic influence	人为影响
anticline	背斜
anti-clockwise direction	逆时针方向
anti-jamming	抗干扰

anti-spoofing(AS) 反电子欺骗
anti-symmetric 反对称
anti-symmetrical wavelet 反对称小波
any weather condition 全天候
aperture 孔,(照相机)光圈,孔径
aphelion [天]远日点
API 应用程序接口
apogee 远地点
apparent azimuth 视方位角
apparent horizon 视地平线
apparent precession 视岁差,视进动
apparent reflectance 表观反射率
apparent sidereal time 视恒星时
apparent solar day 视太阳日
apparent solar time 视太阳时
apparent time 视时
application programming interface (API) 应用程序接口
applied cartography 实用地图学
applied load 外加荷载
approach 引桥
approaches to the optimal design 优化设计方法
approximate adjustment 近似平差
approximate altitude 近似高度
approximate calculation 近似计算
approximate contour 近似等高线
approximate coordinates 近似坐标

approximate entropy 近似相对熵
approximate formula 近似公式
approximate orientation 概略定向
approximate relief 概略地貌
approximate value 近似值
approximation 近似值
apron 海幔
AR model 自回归模型
arbitrary constant 任意常数
arbitrary origin 任意原点
arbitrary projection 任意投影
arbitrary scale 任意比例尺
arc 弧段
arc length 弧长
arc measurement 弧度测量
ArcGIS 地理信息系统软件 ArcGIS
arch 拱门，弓形结构，拱形
arch bridge 拱桥
archaeological photogrammetry 考古摄影测量
archaeology 考古学
architectural and archaeological Photogrammetry 古建筑与古文物摄影测量
architectural drafting 建筑制图
architectural photogrammetry 建筑摄影测量
architectural plan 建筑平面图
arc-node topology 弧-结点拓扑关系

arctic circle	北极圈
arc-to-chord correction in Gauss projection	高斯投影方向改正
ArcView	地理信息系统软件 ArcView
area based image matching	灰度匹配
area boundary	地区界限
area calculation	面积计算
area sampling	面积采样
area survey	面积测量
area symbol	面状符号
area target	面状目标
argument of a complex number	复数的辐角
argument of latitude	升交角距
arid areas	干旱区
arithmetic	算术
arithmetic coding	算术编码
arithmetic mean	算术平均
arithmetic series	等差级数
ARMA model	自回归滑动平均模型
array	数组,阵列
array camera	阵列摄影机
array optimization	阵列优化
ART	自适应共振理论
artificial ants	人工蚂蚁
artificial earth satellite	人造地球卫星
artificial immune	人工免疫
artificial immune system	人工免疫系统

artificial intelligence(AI)	人工智能
artificial neural network(ANN)	人工神经网络
artificial satellite	人造卫星
artificial target	人工标志点
AS	反电子欺骗
as-built survey	竣工测量
ascending grade	上坡
ascending node	升交点
ascending order	递升序
Ashtech	美国阿什泰克
Asian dust	亚洲沙尘暴
aspect of projection	投影中心
assignment crossover	指派交叉
assumed coordinate system	假定坐标系
assumed mean	假定平均数
assumed value	假定值
assumption	假定,假设
astatic gravimeter	无定向重力仪
astrocompass	天文罗盘
astrodynamics	天文动力学,星际航行动力学
astrogeodetic	天文大地测量的
astrogeodetic datum	天文大地基准
astro-geodetic datum orientation	大地基准的天文大地定位
astro-geodetic deflection of the vertical	天文大地垂线偏差
astro-geodetic leveling	天文大地水准测量

A

astrogeodetic network	天文大地网
astrograph	天体摄影仪
astro-gravimetric leveling	天文重力水准测量
astrolight viewer	星空夜视仪
astrometric orbit	天体测量轨道
astrometry	天体测量学
astronomical almanac, astronomical ephemeris	天文年历
astronomical azimuth	天文方位角
astronomical camera	天文摄影机
astronomical control	天文控制
astronomical coordinates	天文坐标
astronomical determination	天文测定
astronomical geodesy	天文大地测量学
astronomical latitude	天文纬度
astronomical longitude	天文经度
astronomical measurement	天文量测
astronomical meridian	天文子午圈，天文子午线
astronomical observation	天文观测
astronomical orientation	天文定向
astronomical parallel	天文纬圈
astronomical point	天文点
astronomical position	天文位置
astronomical positioning	天文定位
astronomical theodolite	天文经纬仪
astronomical tide	天文潮
astronomical triangle	天文三角形

A

astronomy 天文学
asymmetrical 非对称
asymptotic error 渐近误差
atlas 地图集
atlas type 地图集类型
atmosphere 大气层
atmosphere transfer model 大气辐射传输方程
atmospheric absorption 大气吸收
atmospheric artifact 大气效应
atmospheric correction 大气改正,气象改正
atmospheric correction of sea surface topography 海面地形大气改正
atmospheric dispersion 大气色散
atmospheric drag perturbation 大气阻力摄动
atmospheric effect of gravimeter 重力仪的气压影响
atmospheric effects 大气影响
atmospheric extinction 大气消光
atmospheric fog 大气濛雾
atmospheric noise 大气噪声
atmospheric point spread function 大气点扩散函数
atmospheric pressure 大气压
atmospheric radiative transfer 大气辐射传输
atmospheric refraction 大气折射
atmospheric science 大气科学
atmospheric transmissivity 大气透过率
atmospheric water 大气水汽
atmospheric window 大气窗口
atmospheric zenith delay 大地天顶延迟

atomic clock	原子钟
atomic time	原子时
ATR	自动目标识别
attitude	姿态
attitude and orbit determination	姿态确定和轨道测量
avionics(AODA)	电子设备
attitude deviation	姿态偏离
attitude parameter	姿态参数
attitude-measuring sensor(AMS)	姿态测量遥感器
attraction potential	引力势,引力位
attraction, attractive force	吸引力
attribute	属性
attribute accuracy	属性精度
attribute data	属性数据
attribute data file	属性数据文件
attribute domain	属性域
attribute measure	属性测度
attribute query	属性查询
attribute recognition criterion	属性识别准则
attribute table	属性表
attribute type	属性类型
attribute value	属性值
attributes reduction	知识约简
augmented matrix	增广矩阵
authalic latitude	等积投影纬度
authalic longitude	等积投影经度
authalic projection	等积投影
auto-calibration	自动标定

A

autocollimating eyepiece	自准直目镜
autocorrelation	自相关
autocorrelation function	自相关函数
autocovariance	自协方差
autocovariance function	自协方差函数
automated cartographic generalization	自动制图综合
automated cartographic system	自动制图系统
automated cartography software	自动制图软件
automated cartography system for charting	海图自动制图系统
automated detection	自动探测
automated drafting system	自动绘图系统
automated endmember extraction	端元自动提取
automated mapping	自动制图
automated mapping/facilities management system (AM/FM)	自动制图-设施管理系统
automatic aerial triangulation	自动空中三角测量
automatic alarm control	自动报警控制
automatic cartography	自动化地图制图
automatic chart correction	海图自动改正
automatic clipping	自动剪切
automatic compensator	自动补偿器
automatic coordinate plotter	自动坐标展点仪
automatic data transmission	自动数据传输
automatic detection	自动检测
automatic film advance	自动卷片
automatic identification, automatic recognition	自动识别

automatic level	自动安平水准仪
automatic match	自动匹配
automatic plotter	自动绘图机
automatic plotting	自动绘图
automatic registration	自动配准
automatic search	自动搜索
automatic target recognition (ATR)	自动目标识别
auto-mosaicking	自动镶嵌
autoregressive model(AR model)	自回归模型
autoregressive parameters	自回归参数
autoregressive-moving average model(ARMA model)	自回归滑动平均模型
auxiliary angle	辅助角
auxiliary circle	辅助圆
auxiliary data	辅助数据
auxiliary equation	辅助方程
auxiliary equipment	辅助设备
auxiliary observing vessel	辅助观测船
auxillary contour	间曲线,半距等高线,补充等高线
auxillary traverse	辅助导线
availability	可用性
average draft	平均吃水
average error	平均误差
average error in altitude	高度平均误差
average error in position	位置平均误差
average speed	平均速率

average value	平均值
average value of sound velocity	声速平均值
axiom	公理
axis	轴
axis of symmetry	对称轴
azimuth	方位角
azimuth base-line	方位向基线
azimuth circle	方位圈,地平经圈,方位度盘
azimuth compass	方位罗盘
azimuth determination	方位角测定
azimuth of photograph	像片方位角
azimuthal projection	方位投影

A

B

Baarda's data snooping	巴尔达数据探测
back azimuth	反方位角
back project	反投影变换
back propagation(BP)	反向传播法,逆推学习算法,BP算法
back reflectance	后向反射
back rod	后尺
back substitution	回代
background and target	背景与目标
background estimation	背景估计
background noise	背景噪声
backscatter model	后向散射模型
backscattering coefficient	后向散射系数
backsight(BS)	后视
backup	备份
ballistic camera	弹道摄影机
ballistic photogrammetry	弹道摄影测量
band response function	通道响应函数
band-pass filter	带通滤波器
band-pass filtering	带通滤波
bank paving	护岸,护堤
bar sweeper	硬式扫海具
bare rock	明礁
barograph	气压计

barometer	气压计
barometric leveling	气压高程测量,气压测高
barricade	障碍物,栅栏
Barycentric or dynamical coordinate system	太阳系(力学)坐标系
base angle	底角
base array	换能器阵列
base car park	地下停车场
base map	基础底图
base point	基点
base station	基准站
base/distance ratio	基线距离比
base-band	基带
base-height ratio	基-高比
baseline	基线
baseline of territorial sea	领海基线
baseline adjustment	基线平差
baseline correction	基线改正
baseline direction	基线方向
baseline length	基线长度
baseline measurement	基线测量
baseline network	基线网
baseline point of territorial sea	领海基点
basement	地下室
basic elements of a circular curve	圆曲线要素
basic geodetic survey	基本大地测量
basic network	基本控制网

basic scale	基本比例尺
basic triangulation network	基本三角网
basis	基
basis for rectification	纠正图底
bathymeter	深度计
bathymetric chart	海底地形图
bathymetric sidescan sonar	测深侧扫声呐
bathymetric surveying	海底地形测量
bathymetry	测深
bathyscaph	深海潜水器
battlefield environment simulation	战场环境仿真
bay, gulf, bight	海湾
Bayesian approach	贝叶斯方法
Bayesian classification	贝叶斯分类
Bayesian estimation	贝叶斯估计
Bayesian Network(BN)	贝叶斯网络
Bayes'theorem	贝叶斯定理
beacon	信标
beam	梁，桁条，(光线的)束
beam width	波束宽度
beam-type sea gravimeter	摆杆型海洋重力仪
bearing	方位
bearing accuracy	方向精度
bearing calibration	方位校准
bearing compass	方位罗盘
behavior reaction	行为反应
Beidou	北斗(卫星)

bell-shaped curve	钟形曲线
bench mark description	水准点之记
bench mark elevation	水准点标高
benchmark(BM)	水准基点
berm	路肩,护坡道
Bernese	瑞士伯尔尼大学研发的精密定位与定规软件
Bernoulli distribution	伯努利分布
Bernoulli experiment	伯努利试验
Bessel ellipsoid, Bessel spheroid	贝塞尔椭球
Bessel formula for solution of geodetic problem	贝塞尔大地主题解算公式
best basis	最佳基
best fitting ellipse	椭圆拟合
between-the-lens shutter	中心式快门
bi directional querying	双向搜索
bi directional reflectance	双向反射
bias	偏差
bi-cubic convolution	双三次卷积
bid	标书
bid call	招标
bid opening	开标
bidding	投标
bidirectional reflectance characteristic	二向性反射特性
bidirectional reflectance distribution functions(BRDF)	二向性反射分布函数

B

bidirectional reflectance measurement	二向反射测量
bidirectional reflectance model	二向反射模型
bidirectional reflection	二向性反射
bilinear interpolation	双线性内插
binary image	二值图像
binary system	二进制
binocular	双目镜,双筒望远镜,双目的
binocular sequence image	双目序列影像
binomial	二项式
binomial distribution	二项分布
binomial series	二项级数
binomial theorem	二项式定理
biological vision	生理视觉
biomass	生物量
biomass index transformation	生物量指标变换
biomedical photogrammetry	生物医学摄影测量
bird's eye view map	鸟瞰图
bisection	平分
bisection method	分半法
bisector	平分线,等分线
bit	比特
bit-plane coding	比特平面编码
Bjerhammar problem	布耶哈马问题
BL	建筑红线,建筑界限
blackboard architecture	黑板框架
black-body	黑体

B

blank sheet	空白图幅
blasting	爆破(作业)
blind spot eliminating	盲点消除
blinking method of stere-oscopic viewing	闪闭法立体观察
block adjustment	区域网平差
block adjustment by strips	航线法区域网平差
block diagram	块状图
block figure	图廓数据
block-matching	邻域匹配
blue print	蓝底图
blunder	异常值,粗差
BM	水准基点
body-fixed coordinate system, earth-fixed coordinate system	地固坐标系
Boltzmann's constant	玻尔兹曼常数
border	图廓,边缘,界线
border division	图廓分度
border information	图廓注记
border line	边线,图廓线
borehole	钻孔
borehole position survey	钻孔位置测量
boresight	安置误差
bottom characteristics	底质
bottom characteristics exploration	底质调查
bottom characteristics sampling	底质采样
bottom sampler	底质取样器
bottom sediment chart	底质分布图

bottom sounder	海底地貌探测仪
bottom wire	底索
Bouguer anomaly	布格异常
Bouguer correction	布格改正
boundary	边界
boundary adjustment	边界调整
boundary condition	边界条件
boundary line	界线
boundary mark, boundary point	界址点
boundary pixels removal	边缘点去除
boundary settlement	划定界限
boundary stone	界石
boundary survey	边界测量
bounday demareation	界标
bounded function	有界函数
bounded sequence	有界序列
box classification method	盒式分类法
box dimension	盒维数
BP	反向传播法,逆推学习算法,BP算法
BP ANN	BP神经网络模型
bracket	支架
breakline	转折线
breakthrough accuracy	贯通精度
breakthrough point	贯通点
breakthrough survey	贯通测量
bridge deck	桥面
bridge axis location	桥梁轴线测设

B

bridge construction survey	桥梁施工测量
bridge survey	桥梁测量
bridging of model	模型连接
brightness temperature	亮温
brightness temperature difference	亮温差值
broadcast ephemeris	广播星历
brower	浏览器
Bruns formula	布隆斯公式
Brun's spheroid	布隆斯椭球
brush shade	晕渲
BS	后视
B-tree	二叉树
bubble	水准器,气泡
bubble axis, leveling bubble axis	水准管轴
bubble correction	气泡改正
bubble offset	水准气泡偏差
buffer	缓冲区
buffer analysis	缓冲区分析
bug	故障,缺陷,干扰,雷达位置测定器,窃听器
building area	建筑面积
building axis survey	建筑轴线测量
building cost	建筑费用
building coverage	建筑密度
building density	建筑密度
building design	建筑设计
building detection	建筑物检测

B

building engineering	建筑工程
building engineering survey	建筑工程测量
building extraction	建筑物提取
building facade	建筑立面
building law	建筑条例
building line(BL)	建筑红线,建筑界限
building permit	建设工程规划许可证
building reconstruction	建筑物重建
building site	建筑工地
building size	建筑规模
building subsidence observation	建筑物沉降观测
bundle adjustment	光束法平差
bundle aerial triangulation	光束法空中三角测量
bundle of rays	光束,射束
buoy	浮标
buoy autorecording tide gauge	浮标式自计验潮仪
byte	字节

C

C/A Code	粗码,C/A 码
cable	缆,索
cable-stayed	斜拉桥
CAC	机助地图制图
CAD	计算机辅助设计
cadastral feature	地籍要素
cadastral information system	地籍信息系统
cadastral inventory	地籍调查
cadastral map	地籍图
cadastral mapping	地籍制图
cadastral plan	地籍图
cadastral revision	地籍修测
cadastral survey	地籍测量
cadastral survey management	地籍管理
cadastre	地籍
CAGIS	中国地理信息系统协会
calculations of locations (coordinates)	点位(坐标)计算
calculator	计算器
calculus	微积分学
calendar day	历书日
calibration	检校,校准,标定
calibration accuracy	校准精度

calibration baselines	标定基线
calibration matrix	标定矩阵
calibration of transducer	换能器校准
calibrator	校准器,检定器
CAM	计算机辅助测图/机助测图
camera calibration	摄影机检校
camera calibration	相机标定,摄像机标定
camera of projection	投影仪
camera platform	摄影机平台
camera station, exposure station	摄站
campus plan	校园平面图
campus planning	校园规划
canonical correlation	典型相关
canopy model	植物冠层模型
canopy reflectance model	冠层反射模型
cantilever	悬臂
cantilever beam	悬臂梁
capacitor hydrophone	电容水听器
capacity	资源容量
car park	停车场
carrier	载波
carrier phase	载波相位
carrier phase measurement	载波相位测量
carrier-aided tracking	载波辅助跟踪
Cartesian coordiantes	笛卡儿坐标
Cartesian coordinate system	笛卡儿坐标系

C

Cartesian plane	笛卡儿平面
Cartesian product	笛卡儿积
cartogram method	分区统计图法，等值区域法
cartograph	制图仪
cartographer	制图员
cartographic analysis	地图分析
cartographic annotation	地图注记
cartographic classification	地图分类
cartographic communication	地图传输
cartographic database	地图数据库
cartographic editing software	地图编辑软件
cartographic evaluation	地图评价
cartographic exaggeration	制图夸大
cartographic expert system	制图专家系统
cartographic feature	地图地物
cartographic generalization	制图综合
cartographic hierarchy	制图分级
cartographic information	地图信息
cartographic information system (CIS)	地图信息系统
cartographic language	地图语言
cartographic model	地图模型，制图模型
cartographic organization	地图内容结构
cartographic potential information	地图潜信息
cartographic pragmatics	地图语用
cartographic presentation	地图表示法
cartographic selection	制图选取

cartographic semantics	地图语义
cartographic semiology	地图符号学
cartographic simplification	制图简化
cartographic software	地图制图软件
cartographic symbol	制图符号
cartographic syntactics	地图语法
cartography	地图学
cartology	地图学
cartometry	地图量算法
cast-in-place concrete	现浇混凝土
catchment area	汇水面积
catchment area survey	汇水面积测量
category	类型
catenary	悬链
cathode-ray	阴极射线
Cauchy principal value	柯西主值
Cauchy sequence	柯西序列
Cauchy-Schwarz inequality	柯西-许瓦尔兹不等式
caution sign	警示标志
CCD	电荷耦合器件
CCD Array Scanners	CCD 面阵扫描器
CCD camera(charge-coupled device camera)	CCD 摄影机
CCD imagery	CCD 影像
CCD Line Scanners	CCD 线阵扫描器
CCD push-broom image	CCD 推扫影像
celestial body, celestial object	天体

C

celestial coordinate	天球坐标
celestial coordinate system	天球坐标系
celestial equator	天球赤道
celestial map	天体图
celestial mechanies	天体力学
celestial meridian	天球子午线
celestial parallel	天球纬圈
celestial pole	天极
celestial sphere	天球
cell	格网单元
cell size	格网单元尺寸
cement	水泥
center line survey, location of route	中线测量
center of sheet	图幅中心
centering error	对中误差
centering of bubble	气泡居中
centering of instrument	仪器对中
centering rod	对中杆
centimeter	厘米
central city	中心城市
central cross-hair	中丝
central line	中线
central meridian	中央子午线
central projection	中心投影
centre of earth	地心
centre of gravity	重心
centre of rotation	旋转中心
centreline of the road or railway	道路(铁路)中线

C

centrifugal force	离心力
centrifugal potential	离心力位
centripedal acceleration	向心加速度
centripetal force	向心力
centroid	质心
CERS	城市应急联动系统
certain event	必然事件
chain code	链码
chain network	三角锁网
CHAMP	重力卫星
change detection	变化检测
Chang-E satellite	嫦娥卫星
channel	航道,通道
chaos immune algorithm	混沌免疫算法
character merger	特征融合
characteristic coding	特征编码
characteristic curve of photographic emulsion	感光特性曲线
characteristic equation	特征方程
characteristic function	特征函数
characteristic level of water	特征水位
characteristic matrix	特征矩阵
characteristic product	特征量积
characteristic root	特征根
characteristic segmentation	特征分割
characteristics atmospheric trans-missivity	大气传输特性
charge coupled device(CCD)	电荷耦合器件

C

charge injection device(CID)	电荷注入器件
chart	海图
chart accuracy	海图精度
chart atlas	海图集
chart boarder	海图图廓
chart compilation	海图编制
chart content	海图内容
chart correction	海图改正
chart current meter	印刷海流计
chart data base	海图数据库
chart datum	海图基准面
chart large correction	海图大改正
chart lettering	海图标记
chart numbering	海图编号
chart of bottom quality	海底底质图
chart projection	海图投影
chart reproduction	海图制印
chart revision, chart update	海图更新
chart scale	海图比例尺
chart scribing	海图刻图
chart sheet	地图图幅
chart small correction	海图小改正
Chart Standardization Committee of IHO	国际海道测量组织海图标准化委员会
chart subdivision	海图分幅
chart symbols	海图符号
chart title	海图标题
charted depth	图载水深

C

charting	海图制图
check board	检查板
check computation	核算
check line of sounding	检查测深线
check station	监视台,检查台
checking	验算
China Association for Geographic Information System(CAGIS)	中国地理信息系统协会
China Center for Resources Satellite Data and Application(CRESDA)	中国资源卫星应用中心
China Interagency Coordinating Committee on Geo-spatial Data	中国地理空间信息协调委员会
Chinese Society for Geodesy Photogrammetry and Cartography(CSGPC)	中国测绘学会
chord length	弦长
chord offset method	弦线支距法
chorisogram method, cartodiagram method	分区统计图表法
choropleth technique	分级统计图法
choroplethic map	等值区域图
choroplethic method	分区统计图法,等值区域法
chroma	色度
CID	电荷注入器件
CIO	国际协议原点
CIP	瑞士 Zurich 大学研制的 DTM 软件

C

cipher	密码
circle	度盘
circle graduation	度盘分划
circuit closure	环线闭合差
circuit highway	环形公路,环路
circuit railroad	环形铁路
circular bubble	圆水准器
circular curve geometry	圆曲线几何形状
circular curve staking	圆曲线测设
circular encoder	编码度盘
circular error probable (CEP)	圆概率误差
circular function	圆函数,三角函数
circular measure	弧度法
circular motion	圆周运动
circumcentre	外接圆心
circumcircle, circumscribed circle	外接圆
circumference	圆周
circumradius	外接圆半径
CIRS	协议惯性参考系
CIS	地图信息系统
citify	城市化
city agglomeration	城市群
city classification	城市分类
city detailed planning	城市详细规划
city emergency response system (CERS)	城市应急联动系统
city expansion	城市扩张,城市膨胀
city function	城市职能

C

city hall	市政厅
city layout	城市规划,城市布局
city plan	城市平面图
city planning	城市规划
city planning administration	城市规划管理
city planning outline	城市规划纲要
city proper	市区
city regional planning	市域规划
city sitting	城市选址
city size	城市规模
city structure	城市结构
city survey	城市测量
city traffic	市内交通,城市交通,市内运输
city water	自来水
civic landscape	城市景观
civics	市政学
civil architecture	民用建筑
civil building	民用建筑
civil construction	土木工程
civil day	民用日
civil time	民用时
Clairaut theorem	克莱罗定理
clamp	制动
Clarke spheroid	克拉克椭球
class boundary	组界
class interval	组区间
class statistic	聚类统计

classfication rule　分类规则
classical geodetic monitoring techniques　传统大地测量监测技术
classification　分类
classification accuracy　分类精度
classification code　分类码
classification decision　分类决策
classification of charts　海图种类
classification of image texture　影像纹理分类
classified compression　分类压缩
classified enhancement　分类增强
classifier　分类器
clear way　快车道,高速公路
clearance　净空,净度
clearance survey　净空区测量
clearinghouse　数据交换中心
climate change　气候变化
clinometer, inclinometer　倾斜仪
clipping　剪辑
clock bias　钟差
clock frequency　时钟频率
clock offset　钟偏
clock rate　钟速
clock synchronization　时钟同步
clockwise　顺时针
clockwise direction　顺时针方向
clockwise moment　顺时针力矩
clone selection　克隆选择

closed convex region 闭凸区域
closed curve 闭合曲线
closed interval 闭区间
closed leveling line 闭合水准路线
closed loop 闭合环线
closed loop traverse 闭合环导线
closed traverse 闭合导线
close-range photogrammetry 近景摄影测量
closing error 闭合差
closing error in coordinate increment 坐标增量闭合差
closing error of trangle 三角闭合差
closure 闭合差
closure error 闭合差
closure error of azimuth 方位角闭合差
closure error of traverse 导线闭合差
clothoid spiral curve 回旋螺旋曲线
cloud and snow detection 云雪检测
cloud contamination 云干扰
cloud detection 云层检测
cloud-cover 云覆盖
cluster 集群
cluster sample 群样本检验
clustering 聚类
clustering analysis 聚类分析
clustering tree 聚类树
clutter 杂波
coarse adjustment 粗调

coarse sweeping, preliminary sweep	粗扫
Coarse/Acquision Code(C/A Code)	粗码,C/A码
coast line	海岸线
coast line survey	岸线测量
coastal chart	海岸图
coastal current	沿岸流
coastal landform	海岸地貌
coastal survey	沿岸测量,沿海测量
coastal topographic survey	海岸地形测量
coastal water	二类水体
coastal zone	海岸带
coastwise navigation	沿岸航行图
coastwise survey	沿岸测量,沿海测量
coaxial	共轴
coaxial cable	同轴电缆
coaxial circles	共轴圆
coaxial system	共轴系
code correlation technique	码相关技术
code division multiple access (CODE)	码分多址
code of symbols	图符,符号代码
code phase	码相位
coded data	编码数据
coding and decoding	编译码
coding method	编码法
coefficient	系数

C

coefficient of refraction	折射系数
cofactor matrix	协因数矩阵
cofactor matrix of coordinates	坐标协因数矩阵
cofferdam	围堰
cognitive mapping	认知制图
cognitive pattern	认知模式
coherence	相干系数
coherence level	相干度
coherent scattering model	相干散射模型
coincide	重合
coincidence bubble	符合水准器
cold start	冷启动
collimation	照准
collimation adjustment	视准校正
collimation axis	视准轴
collimation error	视准差
collimation line method	视准线法
collimation mark	框标
collimation point	框标点
collinear planes	共线面
collinearity	共线
collinearity equation	共线方程
collocation	配置
color aerial image	彩色航空影像
color chart	多色海图
color chart	地图色标
color coding	彩色编码
color coordinate system	彩色坐标系

color distortion	颜色扭曲
color distortion correction of image	影像色彩失真校正
color enhancement	彩色增强
color film	彩色片
color infrared film, false color film	彩色红外片,假彩色片
color management system	色彩管理系统
color manuscript	彩色样图
color matching	色彩匹配
color meter	水色计
color moments	颜色矩
color of water	水色
color photography	彩色摄影
color proof	彩色校样
color reproduction	彩色复制
color scanner	电子分色机
color separation	分色,分色参考图
color space	颜色空间
color space transformation	颜色空间转换
color transformation	彩色变换
color wheel	色环
colour absorber	滤光镜,滤光片,消色器
column matrix, column vector	列矩阵,列向量
column vector	列向量
columnar water vapour content	水汽总量
combination	组合
combination classification	组合分类

C

combined bundle block adjustment	联合光束法平差
combined (simultaneous) adjustment	联合平差
command tracking station (CTS)	指令跟踪站
Committee on the Exchange of Digital Data	国际海道测量组织数字数据交换委员会
common chord	公弦
common difference	公差
common logarithm	常用对数
common tangent	公切
communication control unit	通信控制器
communication device of water level	水位遥报仪
community structure	群落结构
commutative law	可交换律
comparative cartography	比较地图学
comparator	坐标量测仪,比长仪,检定器
comparison point	比对点
comparison with adjacent chart	邻图对接对比
compass	罗盘
compass bearing	罗盘方位角
compass declination, magnetic declination	磁偏角
compass error	罗盘误差
compass survey	罗盘仪测量
compass theodolite	罗盘经纬仪
compass traverse	罗盘仪导线

C

compensated geoid 补偿大地水准面
compensation current 补偿流
compensation error 补偿器补偿误差
compensator 补偿器
compensator level 自动安平水准仪
compilation 编绘
compiled map 编制图
compiler 编绘员
complementary angle 余角
complementary event 互补事件
complementary function 余函数
complementary probability 互补概率
complementary relationship 互补相关
complete oscillation 全振动
complex conjugate 复共轭
complex conjugate 共轭复数
complex correlation function 复相关函数
complex feature 复杂要素
complex number 复数
complex number plane 复数平面
complex object 复杂目标
complex polygon 复杂多边形
complex root 复数根
complex wavelet transform 复数小波变换
component 分量
component analysis 成分分析
component GIS 组件式地理信息系统
component of force 分力

C

component temperature 组分温度
composite drawing chart 海图编绘
composite function 复合函数
composite number 复合数
composite sailing 组合航法
composite sampling 混合采样
compound angle 复角
compound curve 复曲线
compound probability 合成概率
compound tide 混合潮
comprehensive atlas 综合地图集
comprehensive map 综合地图
comprehensive planning 综合规划
comprehensive planning 总平面图，总体规划
compressed domain 压缩域
computed tomography(CT) 计算机体断层成像
computer aided design (CAD) 计算机辅助设计
computer aided mapping (CAM) 计算机辅助测图/机助测图
computer cartographic generalization 计算机制图综合
computer cartography 计算机地图制图
computer compatible tape (CCT) 计算机兼容磁带
computer graphics 计算机图形学
computer mapping 计算机制图
computer science 计算机科学
computer tomography scanner 计算机断层扫描仪
computer vision 计算机视觉

C

computer-aided cartography (CAC), computer-assisted cartography(CAC)	机助地图制图
computer-assisted classification	机助分类
computer-assisted mapping, computer-assisted plotting, computer-aided mapping	机助测图
concave	凹
concave down	凹向下的
concave polygon	凹多边形
concave up	凹向上的
concentric	共圆
concentric circles	同心圆
conceptual data model	概念模型
conclusion	结论
concrete	混凝土
concrete lining	混凝土衬砌
concrete pillar	混凝土标石
concrete post	混凝土标石
condition adjustment	条件平差
condition adjustment with parameters	附参数条件平差
condition equation	条件方程
condition of closure	闭合条件
condition of intersection	交线条件
conditional identity	条件恒等式
conditional inequality	条件不等式
conditional probability	条件概率

C

conditioned pattern spectrum	条件模式谱
conductivity sensor	电导率传感器
confidence	置信度
confidence ellipse	置信椭圆
confidence flag	置信度
confidence interval	置信区间
confidence level	置信水平
confidence limit	置信极限
configuration problem	图形(结构)问题
conformal longitude	等角投影经度,正形投影经度
conformal projection	等角条件,正形投影
Conformity_congregate Information Field(CCIF)	综合信息场
congruence	全等
congruent figures	全等图形
congruent triangles	全等三角形
conic projection	圆锥投影
conjugate axis	共轭轴
conjugate diameters	共轭(直)径
conjugate hyperbola	共轭双曲
conjugate ray, corresponding image rays	同名光线,共轭射线
conjugation	共轭
connecting traverse	附合导线
connecting triangle	联系三角形
connecting triangle method	联系三角形法
connection survey	联系测量

C

connection survey in mining pit	采区联系测量
connectivity	连通性
connector	接头,插头,转接器
constant	常数
constant error	常差
constant force	恒力
constant of instrument	仪器常数
constant speed	恒速率
constant term	常数项
constant velocity	恒速
constellation	星座
constrained optimization	约束最优化
constraint	约束
constraint condition	约束条件
construction specification	施工规范
construction control network	施工控制网
construction crew	施工队
construction cycle	施工周期
construction detail	施工详图
construction layout	施工放样
construction map	建筑地图
construction material	建筑材料
construction plan	建筑平面图
construction sequence	施工顺序
construction shadow	建筑物阴影
construction stage	施工阶段
construction supervision	施工监督
construction survey	施工测量

construction tolerance	容许施工限差
contact printing	接触晒印
contact screen	接触网屏
continental island	大陆岛
continental margin	大陆边缘
continental rise	大陆隆
continental shelf	大陆架
continental shelf bathymetric chart	大陆架地形图
continental shelf topographic survey	大陆架地形测量
continental slope	大陆坡
continuity	连续性
continuous data	连续数据
continuous function	连续函数
continuous observation	连续观测
continuous strip	连续航带
continuous tone	连续调
continuous tracking	连续跟踪
continuously operating reference system(CORS)	连续运行参考站系统
contional distribution	条件分布
contional expectation	条件期望
contional probilitity	条件概率
contour	等高线,轮廓线
contour accuracy	等高线精度
contour drafting, contour drawing	等高线绘制
contour extraction	轮廓提取
contour interval	等高距
contour label	等高线标注

contour map	等高线图
contour plane	等高面
contour tracing	等高线跟踪，轮廓追踪
contrast	反差
contrast coefficient	反差系数
contrast enhancement	反差增强
contrast(CON)	对比度
control area	控制测量区
control network	控制网
control network density	控制网密度
control network for deformation observation	变形观测控制网
control of water level	水位控制
control point	控制点
control room	控制室
control segment	控制部分
control station	控制站
control strip	测控条，骨架航线
control survey	控制测量
controlled mosaic	有控制点镶嵌图，控制镶嵌
controlling depth	可航最浅水深
control-point coordinate(s)	控制点坐标
conventional coordinate system	常用坐标系，惯用坐标系
conventional inertial reference system(CIRS)	协议惯性参考系

C

conventional international origin (CIO)	国际协议原点
conventional method	常规法
conventional name	惯用名
conventional terrestrial reference system(CTRS)	协议地球参考系
convergence	收敛
convergence of meridine	子午线收敛角
convergence point	收敛点
convergent	收敛的
convergent photography	交向摄影
convergent series	收敛级数
conversion	换算,转换
conversion factor	换算因子
converter	转换器,交换器,换能器,变频管,变频器,转换反应堆
conveyance tunnel	运输隧道
convolution	卷积
co-occurrence matrix	共生矩阵
coordinate	坐标
coordinate adjustment	坐标平差
coordinate axis	坐标轴
coordinate cadastre	坐标地籍
coordinate conversion	坐标变换
coordinate difference	坐标差
coordinate increment	坐标增量
coordinate method	坐标法

coordinate system	坐标系
coordinate transformation	坐标转换
coordinatograph	坐标仪,直角坐标展点仪,绘图仪
coplanar	共面
coplanar conditions	共面条件
coplanar forces	共面力
coplanar lines	共面
coplanarity equation	共面方程
coriolis force	柯氏力
corner detector / detection	角点检测
correction for aberration	像差改正
correction for centering	归心改正
correction for deflection of the vertical	垂线偏差改正
correction for deformation	变形改正
correction for direction	方向改正
correction for index error	指标差改正
correction for meridian curvature	子午线曲率改正
correction for parallax	视差改正
correction for scale distortion	纵横比例校正
correction for tilt angle	倾角改正
correction for zero drift	零点漂移改正
correction of atmospheric effects	大气影响校正
correction of sounder	测深仪改正数
correction of soundings	深度改正
correction of tidal zoning	水位分带改正
correction of transducer baseline	换能器基线改正

C

correction of transducer draft	换能器吃水改正
correction of water level	水位改正
correction of zero line	零线改正
correlated observations	相关观测值
correlation	相关
correlation coefficient	相关系数
correlation of active and passive	主被动相关性
correlation of observations	观测值相关性
correlation registration	相关配准
correlation relaxation	相关松弛法
correlator	相关器
corresponding epipolar line	同名核线
corresponding image points	同名像点
cosecant	余割
cosine	余弦
cosine backscatter model	余弦散射模型
cosine formula	余弦公式
cosine method / vector method	余弦法 / 矢量法(基于角锥体原理的方法)
cosmic mapping	宇宙制图
cost	成本
cost function	代价函数
cost reduction	降低成本
cotangent	余切
countable set	可数集
counterclockwise	逆时针
counterclockwise direction	逆时针方向

C

course	路线,路程,航线
course deviation indicator (CDI)	航线偏航指示
course made good (CMG)	从起点到当前位置的方位
course over ground (COG)	对地航向
course plotting	航迹绘算
course to steer(CTS)	到目的地的最佳行驶方向
course triangle	航行三角形
covariance function	协方差函数
covariance matrix of the observables	观测值协方差矩阵
coverage	图层
coveyer belt	传送带
crack monitoring pins	裂缝监测仪
CRESDA	中国资源卫星应用中心
crest	山脊
crest line	山脊线
criterion	准则
critical point	临界点
critical region	临界域
critical value	临界值
crop yield prediction	农作物单产
cross calibration	交叉定标
cross correlations	互相关
cross covariance	互协方差
cross level	十字形水准器

cross radiance	交叉辐射
cross section	横断面,断面图,剖面图
cross section area	截面积
cross-coupling effect	交叉耦合效应
crosshair, graticule	十字丝,十字瞄准线
crossing	交叉路
crossover adjustment	交叉点平差
cross-ruling	交叉网线
cross-section drawing	横断面图
cross-section leveling	横断面水准测量
cross-section survey	横断面测量
cross-tile indexing	交叉分区索引
crosswalk	过街人行道
cross-wire micrometer	十字丝测微器
crust deformation measurement	地壳形变观测
crustal deformation	地壳变形
crustal movement	地壳运动
crustal movement observation network	地壳运动观测网
crystal eyeglasses	液晶眼镜
crystal shaded eyeglasses	液晶遮光眼镜
CSGPC	中国测绘学会
CT	计算机断层成像
CTRS	协议地球参考系
cubic	三次方
cubic equation	三次方程
cubic meter (m)	立方米

cubic millimeter (mm)	立方毫米
cubic root	立方根
cultural feature	人工地物,人文要素
culture symbol	地物符号,人工建筑符号
cumulative space covering ratio	累积空间覆盖率
current chart	海流图
current information	现势资料
current measurement buoy	测流浮标
current meter	海流计
current observation	海流观测
current theory	海流学
current velocity	流速
curvature	曲率
curvature correction	曲率改正
curvature of earth	地球曲率
curvature of parallel	平行圈曲率
curvature radius	曲率半径
curve	曲线
curve fitting (approximation)	曲线拟合(近似)
curve layout	曲线放样
curve of water level	水位曲线
curve sketching	曲线描绘(法)
curve tracing	曲线描迹(法)
curved surface	曲面
curved surface area	曲面面积
curved surface fitting	曲面拟合
curved tunnel	曲线隧道

cut and cover method	明挖回填法
cut and fill balance	填挖方平衡
cut and fill estimate	土方计算
cut-fill volume	填挖方量
cutoff angle, mask angle	截止高度角
cuts off	截尾
cutting head	刃首
cybercity	数码城市
cyberspace	赛博空间
cycle slip	周跳
cycle slip detection and repair	周跳探测与修复
cycloid	摆线
cylinder	圆柱体
cylindrical	圆柱形的
cylindrical coordinate	柱面坐标
cylindrical harmonics	圆柱调和函数，贝塞尔函数
cylindrical projection	圆柱投影
czapski condition	交线条件

D

daily mean sea level	日平均海面
dam construction survey	大坝施工测量
dam deformation observation	大坝变形观测
dam monitoring	大坝监测
dam site investigation	坝址勘察
DAMCS (Digital Automatic Map Compilation System)	全数字化自动测图系统(第一套数字摄影测量系统,19世纪60年代在美国生产)
damped oscillation	阻尼振动
danamical system	动态系统
danger sounding	危险水深
dangerous gases	有毒气体
dark tone	暗色调,深色调
dasymetric map	分区密度地图
data accessibility	数据的可得性
data accuracy	数据精度
data acquisition	数据获取
data analysis and interpretation	数据分析与解释
data archiving	数据归档
data base	数据库
data capture	数据采集
data classification	数据分类
data cleaning	数据清理

data collection	数据搜集
data collector	数据采集器
data compatibility	数据相容性
data completeness	数据完全性
data compression	数据压缩
data consistency	数据一致性
data conversion	数据转换
data currency	数据现势性
data dictionary	数据字典
data distribution	数据分发
data download	数据下载
data editing	数据编辑
data encoding	数据编码
data exchange format(DXF)	数据交换格式
data fitting	数据拟合
data format	数据格式
data fusion	数据融合
data integration	数据集成
data integrity	数据完整性
data layering	数据分层
data lineage	数据志
data maintenance	数据维护
data management	数据管理
data mining	数据挖掘
data model	数据模型
data normalization	数据正规化
data of inner orientation	内方位元素
data processing	数据处理

data processing system	数据处理系统
data protection	数据保护
data quality	数据质量
data quality control	数据质量控制
data reality	数据真实性
data recorder	电子手簿，数据采集器
data reduction	数据缩减
data reduction, data compression	数据压缩
data retrieval	数据检索
data revision, data update	数据更新
data sampler	数据样品
data sampling rate	数据采样率
data set	数据集
data sharing	数据共享
data simplification	数据简化
data snooping	数据探测法
data sources	数据源
data standard	数据标准
data structure	数据结构
data substitution	资料替补
data transfer	数据转换
data transmission	数据传输
data type	数据类型
data vectorization	数据矢量化
data visualization	数据可视化
data warehouse	数据仓库
data window	数据窗

database architecture, database structure	数据库结构
database design	数据库设计
database for urban survey	城市测量数据库
database management	数据库管理
database management system	数据库管理系统
database manager	数据库管理员
database system	数据库系统
datalogger	数据记录设备
datum	基准面
datum error	基准误差
datum line	基准线
datum of soundings	测深基准面
datum point	基准点
DCCS(Digital Comparator Correlation System)	海拉瓦的混合数字摄影测量系统
DDN	数字数据网
dead reckoning	船位推算
decameter	十米
decay constant	衰变常数,裂变常数
decay curve	衰变曲线,衰减曲线
decelaration	减速度
decentering distortion	偏心畸变差
decimal	小数
decimal point	小数点
decimeter	分米
decision box	判定框
decision making	决策

decision-making tree 决策树
declination arc 磁偏角弧
declination axis 经纬轴
declination circle 经纬圈
declination constant 磁偏常数
declination station 磁偏点
decoding 译码,解码
decomposable Markov network (DMN) 马尔科夫随机网
decomposing method of mixed pixel 混合像元分解
decomposition and reconstruction 分解与重建
decomposition technique 分解技术
deconvolution 退卷积
decrease 递减
decreasing function 递减函数
decreasing series 递减级数
decrement 减量
deduce 演绎
deduction, inference 推论
deductive reasoning 演绎推理
deep space detection 深空探测
definite integral 定积分
deflection angle 偏角法
deflection observation 挠度观测
deflection of the vertical 垂线偏差
deformable body 变形体
deformation 变形,畸变
deformation analysis 变形分析

deformation interpretation 变形解释
deformation monitoring (observation) 变形监测(观测)
deformation observation control network 变形观测控制网
deformation parameter 变形参数
deformation vector 变形向量
degenerated conic section 降级锥曲线
degrees of freedom 自由度
delay lock loop(DLL) 延迟锁相环
DEM 数字高程模型
demodulation 解调
demographic data 人口统计数据
demographic database 人口统计数据库
denoising 去噪
denominator 分母
densification 加密
densification network 加密[控制]网
densitometer 密度计
density 密度
density analysis 密度分析
density current 密度流
density detection 密度检测
density of local sea water 海水现场密度
density of projected points 投影点密度
density of sea water 海水密度
density slicing 密度分割
depression angle 俯角

depressor	阻浮器
depth	深度
depth datum	深度基准面
depth instrument	测深仪器
depth of field	景深
depth perception	深度感
depth sounder	测深仪
depth sounding	声波探测
depth tracing	深度透写图
depth wire	深度索
derivative	导数
descending order	递降序
desertification	荒漠化
design elevation	设计高程
design matrix	设计矩阵
design paper	设计图纸
design parameter	设计参数
design phase	设计阶段
design water level	设计水位
designated function of city	城市性质
designed breakthrough accuracy	设计贯通精度
designed sound velocity of sounder	测深仪设计声速
desired track (DTK)	期望航线(从起点到终点的路线)
destriping	条带去除
detached building	孤立建筑物
detail design	详细设计
detail drawing	详图,细部图

detail point	碎部点
detail survey	碎部测量
detail work drawing	施工详图
detailed planning	详细规划
detailed triangulation	图根三角测量,低等三角测量
detected signal	探测信号
detecting model	探测模式
detection	检测
detection of outliers	粗差探测
detection template	检测模板
detector	探测器,探元
determinant	行列式
determination of longitude	经度测定
determination of map size	地图尺寸测定
determination of position	位置测定
determination of scale	比例尺测定
determination of tilt	倾角测定
deterministic function	确定性函数
development area	开发区
development of land	土地开发
deviation	偏差
deviation from the mean	离均差
DFT	离散傅立叶变换
DG direct georeferencing	直接对地目标定位
DGM	数字地面模型
DGPS	差分 GPS
DHM	数字高程模型

diagonal	对角线
diagonal element	对角线元素
diagonal matrix	对角阵
diagram	图表
diapositive	透明正片
diazo copying	重氮复印
dielectric constant	介电常数
difference	差
difference equation	差分方程
difference of elevation	高程差
difference of height	高差
difference of latitude	纬差
difference of longitude	经差
differential	微分
differential correction	差分改正
differential equation	微分方程
differential GPS (DGPS)	差分 GPS
differential interferometry	差分干涉测量法
differential leveling	微差水准测量
differential mean value theorem	微分中值定理
differential positioning	差分定位
differential rectification	微分纠正
differentiation	微分法
diffractive optics	二元光学元件
diffuse reflectance	漫反射
diffusion coefficient	扩散系数
diffusion transfer	扩散转印
DIGEST	数字地理信息交换标准

D

digit	数字
digital camera	数码相机
digital cartographic data standard	数字制图数据标准
digital cartography	数字地图学
digital China	数字中国
digital city	数字城市
digital data network(DDN)	数字数据网
digital distortion model(DDM)	数字畸变模型
digital dual-frequency echosounder	数字双频回声测深仪
digital earth	数字地球
digital elevation model (DEM), digital height model (DHM), digital terrain elevation model (DTEM)	数字高程模型
digital file	数字化文件
digital filter	数字滤波器
digital filtering	数字滤波
digital geographic information exchange standard(DIGEST)	数字地理信息交换标准
digital image	数字影像
digital image processing	数字图像处理
digital landscape model(DLM)	数字景观模型
digital line graph(DLG)	数字线画地图
digital map	数字地图
digital mapping	数字测图
digital mosaic	数字镶嵌
digital orthoimage	数字正射影像
digital orthophoto map (DOM)	数字正射影像图
digital photogrammetric work station	数字摄影测量工作站

D

digital photogrammetry	数字摄影测量
digital plotter	数控绘图机
digital raster graphics(DRG)	数字栅格图
digital reading	数字读数
digital rectification	数字纠正
digital seismometer	数字地震仪
digital signal generator	数字信号源
digital surface model(DSM)	数字表面模型
digital tape	数字磁带
digital terrain model(DTM), digital ground model (DGM)	数字地面模型
digital tracing table	数控绘图桌
digital watermarking	数字水印
digitization	数字化
digitized image	数字化影像
digitized map	数字化地图
digitizer	数字化仪
digitizing accuracy	数字化精度
dilatation	膨胀
dilution of precision (DOP)	精度因子
dimension	维(数)
dimension of a matrix	矩阵的阶
DInSAR	差分合成孔径雷达干涉测量
DIPS	德国、瑞士、苏黎世合作研制的实时摄影测量系统
direct (mathematical) solution	直接(解析)法

D

direct adjustment	直接平差
direct leveling, spirit leveling	几何水准测量
direct linear transformation(DLT)	直接线性变换
direct plummet observation	正锤[线]观测
direct radiance	直射辐射
direct scheme of digital rectification	直接法纠正
direct solution of geodetic problem	大地主题正解
direction	方向
direction angle	方向角
direction calculation	方向计算
direction checking	方向检测
direction cosine	方向余弦
direction of gravity	重力方向
direction of plumb-line	垂线方向
direction of precession	进动方向
direction of spin	旋转方向
direction of tilt	倾角方向
direction ratio	方向比
directional antenna	定向天线
directional brightness temperature	方向亮温
directional thermal emission	热辐射方向性
disaster management	灾害管理
disaster prevention	防灾
discharge structure	泄水建筑物
discontinuity	不连续性
discontinuous	间断(的)
discontinuous point	不连续点

discrete　离散
discrete control system　离散控制系统
discrete cosine transform(DCT)　离散余弦变换
discrete data　离散数据
discrete Fourier transform(DFT)　离散傅立叶变换
discrete random variable　离散随机变量
discrete uniform distribution　离散均匀分布
discriminant　判别式
discriminatory analysis　判别分析
disjoint　不相交的
disjoint sets　不相交的集
disk array　磁盘阵列
dispersion　离差
displacement　位移
displacement observation　位移观测
displacement of image point　像点位移
displacement transducer　位移传感器
disproportional　不成比例的
distance　距离
distance accuracy　测距精度
distance calculation　距离计算
distance check　距离检核
distance correction in Gauss projection　高斯投影距离改正
distance decision function　距离判决函数
distance education　远程教育
distance formula　距离公式
distance measurement　测距

D

distance measuring instrument	测距仪
distance root mean square (error)	距离均方根(误差)
distance run	航程
distance-based matching	距离匹配
distance-measuring error	测距误差
distinct roots	相异根
distinct solution	相异解
distorted pixel correction	像元畸变校正
distortion	畸变
distortion isograms	等变形线
distortion error	畸变差
distribution	分布
distribution function	分布函数
distribution table	分位表
distribution value	分位值
distributive law	分配律
district planning	分区规划
disturbing potential	异常位,扰动位
dithering	抖动
diurnal inequality	日不等
diurnal motion	周日运动
diurnal parallax	周日视差
diurnal tidal current	全日潮流
diurnal tide harbor	日潮港
diurnal variation	日变
divergence	发散(性)
divergent	发散的
divergent iteration	发散性迭代

divergent series	发散级数
divided circle	刻度盘
divided scale	分画尺
division	除法
division algorithm	除法算式
DLG	数字线画地图
DLM	数字景观模型
DLT	直接线性变换
dodging	匀光
DOM	数字正射影像图
DOP	精度因子
Doppler Orbit determination and Radiopositioning Integrated on Satellite (DORIS)	多普勒轨道定位与无线电集成系统
Doppler ranging	多普勒测距系统
Doppler shift	多普勒频移
Doppler sonar	多普勒声呐
DORIS	多普勒轨道定位与无线电集成系统
dot method	点值法
dot product	点积
double astrograph	双筒天体摄影仪
double lane	双车道
double leveling, double-run leveling, reciprocal leveling	往返测水准测量
double orientation	双定向
double principle distance	双主距
double root	二重根

double-difference	双差
doublet transducer	偶极子换能器
DPW (Digital Photogrammetry Workstation)	美国 LH Systems 的卫星遥感测图处理系统
draft	吃水,草图,设计图样
draft correction	吃水改正
draft performance standard	国际海道测量组织制定的电子海图显示与信息系统性能标准草案
draft plan	草图
drafting accuracy	绘图精度
drag sweep	拖底扫海
drag wire	拖索
draignage system	排水系统
drain pipe	排水管
drainage design	排水设计
drainage map	水系图
drawing office	设计室,制图室
DRG	数字栅格图
drill & blast method	钻孔爆破法
drilling	钻孔
drilling machine	钻孔机
dropping resistors	减压电阻器,降压电阻器
drought	干旱
drought prediction model	干旱预警模型
drum scanner	鼓扫描器

dry land	旱地
drying rock	干出滩
DSCC(Digital Stereo Compatator/Compiler)	美国国防测图局的卫星遥感测图处理系统
DSM	数字表面模型
DSP1 (Digital Stereo Photogrammetric System)	数字立体摄影测量系统(瑞士 Kern 与英国剑桥 GEMS 公司共同研制,1988 年 7 月推出)
DTEM	数字高程模型
DTM	数字地面模型
dual polarization	双线偏振
dual stereo synchro photography	双立体同步摄影
dual-frequency	双频
dual-frequency sounder	双频测深仪
duality	二元性
dust detection	沙尘监测
dust source area	沙尘源区
DVP(Digital Video Plotter)	加拿大的卫星遥感测图处理系统
DXF	数据交换格式
dynamic change	动态变化
dynamic correction	动力高改正
dynamic draft	动吃水
dynamic ellipticity of the earth	地球动力扁率
dynamic factor of the earth	地球动力因子
dynamic geodesy	动力大地测量学

dynamic GIS	动态地理信息系统
dynamic height	力高
dynamic hienchy discriminatory analysis	动态综合层次判别分析
dynamic landscape simulation	动态地景仿真
dynamic load	动荷载
dynamic monitoring	动态监测
dynamic ocean topography	动力海面地形
dynamic prediction	动态预测
dynamic programming	动态规划
dynamic sensor	动态传感器
dynamic system	动态系统
dynamic variable	动态变量
dynamic window	动态窗口
dynamical evolutionary algorithm	动力演化算法
dynamical oceanography	动力海洋学
dynamical parallax	动力视差
dynamical theory of tides	潮汐动力论
dynamics	动力学

E

Earth axis	地轴
Earth centered Earth fixed	地心地固
earth dam	土坝,土堤
Earth ellipsoid, Earth spheroid	地球椭球
earth embankment	土堤
Earth gravity model	地球重力场模型
Earth observing satellites	对地观测卫星
Earth observing system (EOS)	对地观测体系
Earth orientation parameter(EOP)	地球定向参数
Earth resources technology satellite (ERTS)	地球资源卫星
Earth sciences	地球科学
Earth shape	地球形状
Earth tidal parameters	地球潮汐参数
Earth tide	地球潮汐,固体潮
Earth tide observation	固体潮观测
Earth wobbles	地球摆动
Earth's (forced) nutation	地球章动
Earth's annual motion (revolution about the sun)	地球周年视运动(绕太阳旋转)
Earth's diurnal motion (spin)	地球周日视运动(自转)
Earth's flattening	地球扁率
Earth's free core nutation(FCN)	地球自由核章动

Earth's oscillation	地球振荡
Earth's precession	地球岁差进动
Earth's solid inner core	地球固体内核
Earth's spin	地球旋转
earthquake	地震
earthquake intensity	地震强度
earthquake magnitude	地震震级
earthquake origin	地震震源
earthquake-proof construction	抗震建筑
earth-retaining wall	挡土墙，护土墙
Earth's rotation	地球自转
Earth's rotation parameters (ERP)	地球自转参数
earthwork volumes	土方量
easement curve	缓和曲线
East Asia Hydrographic Commission	国际海道测量组织东亚海道测量委员会
East Atlantic Hydrographic Commission	国际海道测量组织东大洋海洋测量委员会
east longitude	东经
easting	东距
ebb current	落潮流
ebb tide	落潮
eccentric angle	偏心角
eccentric circles	离心圆
eccentric reduction	归心计算
eccentricity	偏心率，离心率
eccentricity correction	偏心改正
eccentricity of ellipsoid	椭球偏心率

E

ECDB	电子海图数据库
ECDIS	电子海图显示和信息系统
echelon form	梯阵式
echelon matrix	梯矩阵
echo altimeter	回波测高仪
echo signal	回波信号
echo signal of sounder	测深仪回波信号
echo sounder	回声测声器,回声探测器,回声测深仪
echo sounding	回声测深
echo sweeper	回声扫测仪
echogram	测深仪记录
eclipse	日蚀,月蚀,蒙蔽
ecliptic	黄道
ecliptic latitude	黄纬
ecliptic longitude	黄经
ecliptic pole	黄极
ecological balance	生态平衡
ecological change	生态变化
ecological city	生态城市
ecology	生态学
economic map	经济地图
economic planning	经济规划
economic status	经济地位,经济状态
ecosphere	生态圈
ecosystem	生态系统
edge	边缘

E

edge detection 边缘检测
edge enhancement 边缘增强
edge extraction 边缘提取
edge keeping 边缘保护
edge linking 边缘连接
edge matching 图幅接边
edge operator 边缘算子
edge-based matching 基于边缘的匹配
EDM 电子测距仪
eduacational facilities 教育设施
effect of atmospheric downward thermal radiance 大气下行辐射效应
effective emissivity 有效发射率
effective radius 有效半径
effective range 有效范围
efficient estimator 有效估计量
EGNOS 欧洲静地导航覆盖服务
eigenvalue 本征值,特征值
eigenvector 本征向量,特征向量
Ekman current meter 厄克曼海流计
elastic body 弹性体
elastic collision 弹性碰撞
elastic constant 弹性常数
elastic deformation 弹性变形
elastic failure 弹性破坏
elastic force 弹力
elastic settlement 弹性沉降

elasticity	弹性
electric conduction	导电
electric field	电场
electric shaft	电气竖井
electrode polarization	电极极化
electromagnetic distance measurement	电磁波测距
electromagnetic distance measuring instrument	电磁波测距仪
electromagnetic log, EM log	磁计程仪
electromagnetic radiation	电磁辐射
electromagnetic spectrum	电磁波频谱,电磁波谱,电磁光谱
electromagnetic spectrum energy	电磁波谱能
electronic atlas	电子地图集
electronic chart	电子海图
electronic chart database(ECDB)	电子海图数据库
electronic chart display and information system(ECDIS)	电子海图显示和信息系统
electronic conductor	电子导电体
electronic correlation	电子相关
electronic distance measuring instrument(EDM)	电子测距仪
electronic hydrophone	电子水听器
electronic level	电子水准仪
electronic map	电子地图
electronic navigation chart database (ENCDB)	电子导航海图数据库

E

electronic plane-table	电子平板仪
electronic planimeter	电子求积仪
electronic publishing system(EPS)	电子出版系统
electronic scanner	电子扫描仪
electronic scanning	电子扫描
electronic sensing device	电子传感器
electronic theodolite	电子经纬仪
electro-optical distance measuring instrument	光电测距仪
element	元素
element of rectification	纠正元素
element of relative orientation	相对定向元素
elementary event	基本事件
elementary function	初等函数
elementary matrix	初等矩阵
elements of absolute orientation	绝对定向元素
elements of centring	归心元素
elements of exterior orientation	像片外方位元素
elements of interior orientation	像片内方位元素
elevation accuracy	高程精度
elevation angle	高度角
elevation calculation	高程计算
elevation circle	垂直度盘,竖盘
elevation control	高程控制
elevation difference	高差
elevation notation	高程注记
elevation of sight	视线高
elevation point	高程点

elevation, height	高程
elimination	消元
elimination method	消元法
ellipse	椭圆
ellipsoid	椭球,椭球体
ellipsoid of revolution	旋转椭球体
ellipsoidal coordinates	椭球体面坐标
ellipsoidal curvature	椭球面曲率
ellipsoidal distance	椭球体面距离
ellipsoidal harmonics	椭球调和函数,椭球谐函数
ellipsoidal height	大地高
ellipsoidal meridian	椭球子午线
ellipsoidal normal	椭球法线
elliptic positioning	椭圆定位
elongation	伸张,展
embankment	堤防,筑堤
embankment foundation	堤基
embayed coast	港湾海岸
embedded hidden Markov model (EHMM)	嵌入式隐马尔可夫模型
emergency	突然事件,紧急事件
emergency access	紧急通道
emergency alarm system	紧急报警系统
emergency brake	紧急制动器
emergency channel	紧急频道
emergency control centre	紧急事故控制中心
emergency disposal	应急处置

emergency exercise 应急演习
emergency exit 紧急出口,安全出口,
emergency exit door 紧急出口门
emergency management 应急管理
emergency monitoring and support centre 紧急事故监察及支援中心
emergency plan 应急预案,应急计划
emergency preparedness 应急准备
emergency response 应急
emissivity 发射率
emitter recognition 辐射源识别
empirical data 实验数据
empirical formula 实验公式,经验公式
empirical linear calibration 经验线性定标
empirical mode decomposition (EMD) 经验模态分解
empirical orientation 经验定向,目视定向
empirical probability 经验概率
ENCDB 电子导航海图数据库
enchance 增强
enclosed sea 内陆海
encoding 编码
end point 端点
energy edge 能量边缘
engineering control network 工程控制网
engineering design 工程设计
engineering drafting 工程制图
engineering drawings 工程制图

engineering geological investigation	工程地质勘察
engineering geology	工程地质学
engineering photogrammetry	工程摄影测量
engineering project	工程项目
engineering sciences	工程科学
engineering structures	工程建筑物
engineering survey	工程测量
engineering surveying(geodesy)	工程测量学
engineer's level	工程水准仪
engineer's theodolite	工程经纬仪
entity	实体
entity object	实体对象
entity relationship	实体关系
entropy	熵
entropy coding	熵编码
environmental analysis	环境分析
environmental architecture	环境建筑学
environmental asessment(EA)	环境评估
environmental monitoring	环境监测
environmental planning	环境规划
environmental pollution	环境污染
environmental factors	环境因子
environmental management	环境管理
environmental map	环境地图
environmental survey satellite	环境探测卫星
ENVISAT	微波遥感卫星，2001年发射
EOP	地球定向参数

EOS	对地观测体系
Eotvos correction	厄特弗效应改正
Eotvos effect	厄特弗效应
ephemeris	历书,星历表,星历
ephemeris error	星历误差
ephemeris time	历书时
epicontinental sea	陆缘海
epipolar axis	核轴
epipolar correlation	核线相关
epipolar curve	核曲线
epipolar line,epipolar ray	核线
epipolar plane	核面
epipole	核点
epoch	历元
epoch of partial tide	分潮迟角
EPOS	德国地学研究中心研发的精密定位与定轨软件
EPS	电子出版系统
EQ-90mm-CLR	美国的单面阵航空数码相机
equal area projection	等积投影
equal latitude	等积投影纬度
equal probability	等概率
equal ratios theorem	等比定理
equal sets	等集
equal sighting lengths	等视距
equal value gray scale	等值灰度尺

E

equal-angle projection	等角投影
equal-area projection	等积投影
equality	等(式)
equally tilted photography	等倾摄影
equation	方程
equation of LOP	位置线方程
equation of locus	轨迹方程
equation of motion of the satellite	卫星运动方程
equation of state of sea water	海水状态方程
equator	赤道
equatorial circle	赤道圈
equatorial gravity	赤道重力
equatorial parallax	赤道视差
equatorial plane	赤道面
equatorial radius	赤道半径
equatorial scale	赤道长度比
equidistant	等距(的)
equidistant projection	等距投影
equilateral	等边(的)
equilateral polygon	等边多边形
equilateral triangle	等边三角形
equilibrium	平衡
equilibrium spheroid	均衡椭球
equilibrium temperature	平衡温度
equilibrium tide	平衡潮
equilong circle arc grid	等距圆弧格网
equinoctial colure	二分圈
equipotential surface	等位面

equivalence	等价
equivalence class	等价类
equivalence relation	等价关系
equivalent black-body temperature	等效黑体温度,等价黑体温度
equivalent image	等效影像
equivalent line of annual magnetic variation	周年等磁变线
equivalent projection	等积投影
equivalent time sampling	等效时间采样
equivalent triangle	等积三角形
ERP	地球自转参数
error	误差
error analysis	误差分析
error assessment	误差估计
error distribution	误差分布
error ellipse	误差椭圆
error equation	误差方程
error estimation	误差估算
error in latitude	纬度误差
error in longitude	经度误差
error of closure	闭合差
error of focusing	调焦误差
error propagation	误差传播
error term	误差项
error test	误差检验
error triangle	示误三角形
ERS (Europe remote sensing satellite)	欧洲遥感卫星

E

ERTS 地球资源卫星
ESA 欧空局
estimate 估计,估计量
estimate value 估值
estimated course 推算航向
estimated distance 推算航程
estimated position 推算船位
estimated position error (EPE) 估计位置误差
estimated time of arrival (ETA) 估计到达时间(导航)
estimation 估计,估算
estimation of the spectrum 谱估计
Euclidean algorithm 欧几里得算法
Euclidean geometry 欧几里得几何
Euclidean space 欧几里得空间
Euler's formula 欧拉公式
European Geostationary Navigation Overlay Service(EGNOS) 欧洲静地导航覆盖服务
European Space Agency(ESA) 欧空局
even function 偶函数
even number 偶数
exact solution 精确解
exact value 精确值
excavation 开挖
excavation area 开挖区,开挖面积
excavation guidance 开挖引导
excavation works 挖掘工程
excavator 挖土机
excentricity 偏心常数

excess flow device	溢流装置
exclusive economic zone	专属经济区
exclusive events	互斥事件
existing construction	现有建筑
expanded form	展开式
expectation	期望
expected accuracy	预期精度
expected value	预期值
experimental probability	实验概率
expert knowledge	专家知识
expert system	专家系统
explicit function	显函数
exploration geophysics	勘测地球物理
exploration instrument for geophysics	地球物理勘测仪器
exponential	指数
exponential function	指数函数
exposure	曝光
exposure station	曝光站,摄站
express way	高速公路
extended fractal	扩展分形
extended structure	扩建物
extended target	面目标
extensible markup language(XML)	可扩展置标语言
extensometer	伸缩仪
exterior angle	外角
exterior orientation	外部定向,外定向
exterior orientation element	外方位元素

exterior orientation parameters	外方位元素
external accuracy	外部精度
external bracing	外部支撑
external crack	外表裂缝
external diameter	外直径
external distance	外矢距
external force	外力
external reliability	外部可靠性
extra contour	助曲线,辅助等高线
extraction	提取
extraction of color signal	色彩信号提取
extreme point	极值点
extreme value	极值
extremum	极值
eyepiece adjustment	调焦(检景器目镜的)

F

face left	盘左
face reflectance	前向反射
face right	盘右
factorization of polynomial	多项式因式分解
factory area	工厂区
factory building	厂房
fair drawing	清绘
fairway	航道
false color	假彩色
false color composite	假彩色合成
false color image	假彩色图像
false color photography	假彩色摄影
false corner	伪角点
falsework	脚手架,工作架
family of circles	圆族
family of concentric circles	同心圆族
far range	大测距,远测距
fast approximate principal component analysis algorithm	快速近似主成分分析算法
fast Fourier transform(FFT)	快速傅立叶变换
fast independent component analysis	快速独立分量分析
fast integer ambiguity resolution	快速求解整周模糊度
fast lifting wavelet transform (FLWT)	快速提升小波变换

F

fault 断层,断裂
Faye correction 法伊改正
FCN 地球自由核章动
feasibility study 可行性研究
feasible solution 可行解
feature 要素,特征
feature attribute 要素属性
feature based matching 特征匹配/基于特征的匹配
feature catalogue 要素分类
feature cluster 特征聚类
feature code 特征码
feature code menu 特征码清单
feature coding 特征编码
feature combination 特征组合
feature extraction 特征提取
feature fusion 特征融合
feature identifier 要素标识码
feature points 特征点
feature relationship 要素关系
feature selection 特征选择
feature space 特征空间
feature type 要素类型
Federal Geographic Data Committee, USA, FGDC 美国联邦地理数据委员会
Fermat's last theorem 费尔马最后定理
Ferreros formula 非列罗公式
FFT 快速傅立叶变换

Fibonacci number	斐波那契数,黄金分割数
Fibonacci sequence	斐波那契序列
fictitious force	假想的力,伪力
fiducial mark	框标
fiducial network	基准网
field book	外业手簿
field calculation	现场计算
field check	外业检核
field data	外业数据
field manual	外业规范,外业手册
field map	实测原图
field mapping	野外填图,制图
field of view(FOV)	视场
field procedures	外业步骤
field reconnaissance	野外勘测
field sketch	外业草图
fieldwork	外业
FIG	国际测量师联合会
figure	图(形),数字
figure of the Earth	地球形状
figure-ground discrimination	图形-背景辨别
film flatness	胶片平整度
filter	滤光片,滤光镜
filtering	滤波
final building cost	工程决算
final original	出版原图
final velocity	末速度

fine adjustment	微调,细调,精密平差
fine setting	精确安置
finite dimensional vector space	有限维向量空间
finite element method	有限元法
finite probability space	有限概率空间
finite series	有限级数
finite set	有限集
fire accident	火灾
fire alarm system	火灾报警系统
fire barrier	防火墙
first eccentricity	第一偏心率
first echo (return)	首次回波
first term	首项
first-order accuracy	一等精度
first-order benchmark	一等水准点
first-order design (FOD): configuration problem	一类设计:图形结构设计
first-order geodetic network	一等大地网
first-order leveling	一等水准
first-order leveling network	一等水准网
first-order traverse	一等导线
first-order triangulation	一等三角测量
fish buoy	鱼型浮
fish haven	鱼堰
fish stacks	鱼栅
fishing chart	渔业用图
fishing rock	鱼礁
fissure observation	裂缝观测

F

fitness	适应度
fitting of moving quadric surface	移动曲面拟合法
fixed antenna	固定天线
fixed datum	固定起始值,固定基准
fixed error	固定误差
fixed mean pole	固定平极
fixed phase drift	固定相移
fixed point	固定点
flaps	分版原图
flashing rhythm of light	灯光节奏
flat calibration	平面场定标
flattening of ellipsoid	椭球扁率
flattening of the Earth	地球扁率
flexivity	挠度
flight altitude	飞行高度,航高
flight altitude	航高
flight block	摄影分区
flight line of aerial photography	摄影航线
flight plan of aerial photography	航摄计划
float gauge	浮子验潮仪
floating dot	浮点,浮动测标
floating light	浮动灯标
flood control	洪水控制
flood current	涨潮流
flood disaster	洪涝灾害
flood diversion area	分洪区
flood monitoring	洪水监测

F

flood protection works	防洪工程
flood storage works	蓄洪区
flood tide	涨潮
floor area ration	容积率
floor station	底版测点
floorage	使用面积
floor-on-grade	地坪
flow chart	流程图
flow discharge	流量
fluid outer core	液核
fluorescence line height	基线荧光高度
fluorescence remote sensing	荧光遥感
fluorescent map	荧光地图
flying quality	飞行质量
Flykin	Geosurv 公司 GPS 数据处理软件
f-number, stop-number	光圈号数
focal length	焦距
focal plane shutter, curtain shutter	帘幕快门
focus adjustment	调焦
focus lock	焦距锁定
focusing accuracy	对光精度,调焦精度
focusing error	调焦误差
foot of perpendicular	垂足
footing	基座
footprint	激光脚点
forbidden harmonics	禁用调和函数
forced centering	强制对中

F

fore rod	前尺
fore-and-aft overlap	航向重叠
foresight(FS)	前视
forest land	林地
forest survey	林业测量
formalization description	形式化描述
format conversion	格式转换
formula(formulae)	公式
foul ground	险恶地
foundamental astrometry	基础天体测量学
foundation load	基础荷载
foundation pit	基坑
foundation plan	基础平面图
foundation settlement	基础沉降(陷)
foundation works	基础工程,地基工程
foundation(footing)	基础,地基
four beam sounder	四波束测深系统
four color printing	四色印刷
Fourier transform	傅立叶变换
Fourier-Mellin transform	傅立叶-梅林变换
FOV	视场
FOV effect	视场角效应
fractal	分数维,分形
fractal compression	分形压缩
fractal dimension	分维
fractal geometry	分形几何
fractal measurement	分形测量
fraction	分数

fractional Brownian motion	分形布朗运动
fractional Brownian motion field	分数维布朗运动场
frame camera	框幅摄影机
frame construction	框架结构
frame interline transfer(FIT)	帧-行间传输
frame of reference	参考框架
frame perspective	框幅式
frame transfer	帧传输
frame wavelet	框架小波变换
free fall	自由下坠
free station	自由设站法
free traverse	自由导线,支导线
free vector	自由向量
free-air anomaly	空间异常
free-air correction	空间改正
free-hand sketch	手绘草图
frequency	频率
frequency distribution	频数分布,频率分布
frequency division multiple access (FDMA)	频分多址
frequency domain	频域
frequency drift	频漂
frequency of sounding	测深密度
frequency offset	频偏
frequency response function	频率响应函数
frequency-domain descriptions of time series	时间序列频域描述
friction	摩擦

F

friction coefficient	摩擦系数
fringe	干涉条纹图
from the whole to the part	从整体到局部
FS	前视
fulcrum	支点
full frame transfer	全帧传输
full stereoscopic coverage	全立体覆盖
fully automated	全自动
fully digital mapping	全数字化测图
fully normalized harmonics	完全规范化调和函数,完全规范化谐函数
fully normalized spherical harmonics	完全规范化球谐函数
fully operational capability(FOC)	完全运行能力
fully-polarized	全极化
functional model	函数模型
fundamental frequency	基准频率
fundamental network	基本[控制]网
fusion	融合
fuzzy analysis model	模糊分析模型
fuzzy classification	模糊分类
fuzzy classifier method	模糊分类法
fuzzy clustering	模糊聚类
fuzzy c-means(FCM)	模糊 c-均值
fuzzy compactness	模糊紧支度
fuzzy decision	模糊决策
fuzzy degree	模糊度

fuzzy image	模糊影像
fuzzy logic	模糊逻辑
fuzzy mathematics	模糊数学
fuzzy membership function	模糊隶属度函数
fuzzy neural network	模糊神经网络
fuzzy objects	模糊对象
fuzzy set	模糊集
fuzzy supervised classification	模糊监督分类
fuzzy tolerance	模糊容差

F

G

GA	遗传算法
galaxy	银河系,星系
GALILEO	伽利略系统[欧洲]
GALILEO control centers (GCC)	伽利略控制中心
Galileo positioning system	伽利略卫星定位系统
GAMIT	美国 GPS 精密定位与定轨软件
gap probability	间隙率
GARMIN	GPS 仪器公司
GAST	格林尼治真恒星时
gauge station	水位站
gauss blur	高斯模糊
Gauss plane coordinate	高斯平面坐标
Gauss plane coordinate system	高斯平面坐标系
Gauss projection	高斯投影
Gaussian distribution	高斯分布,正态分布
Gauss-Krüger coordinates	高斯-克吕格坐标
Gauss-Krüger projection	高斯-克吕格投影
Gauss-Markoff model	高斯-马尔柯夫模型
gazetteer	地名录
GDOP	几何精度因子
general astronomy	普通天文学
general atlas	普通地图集
general bathymetric chart of the oceans	大洋地势图

general chart 普通海图
general chart of the sea 海区总图
general form 一般式,通式
general loayout plan 总平面图
general map 总图
general planning 总体规划
general precession 总岁差
general relativity 广义相对论
general solution 通解,一般解
generalized adjustment 广义平差
generalized discriminate analysis 广义判别分析
generalized inverse 广义逆
generalized inverse of a matrix 矩阵广义逆
generalized point photogrammetry 广义点摄影测量
generalized ridge estimate 广义岭估计
generalized spatialized information grid 广义空间信息网格
generating function 母函数,生成函数
generator 母线,生成元
genetic algorithm(GA) 遗传算法
geocenter 地心
geocentric coordinate 地心坐标
geocentric coordinate system 地心坐标系
geocentric datum 地心基准
geocentric geodetic coordinate 地心大地坐标
geocentric gravitational constant 地心引力常数
geocentric latitude 地心纬度
geocentric longitude 地心经度

G

geocentric origin	地心原点
geocentric radius vector	地心向量径
geocentric rectangular coordinate system	地心直角坐标系
geocoding	地理编码
geocoding system	地理编码系统
geodesic	大地线
geodesy	大地测量学
geodetic astronomy	大地天文学
geodetic azimuth	大地方位角
geodetic boundary value problem	大地测量边值问题
geodetic control	大地测量控制
geodetic coordinate	大地坐标
geodetic coordinate system	大地坐标系
geodetic database	大地测量数据库
geodetic datum	大地基准
geodetic gravimetry	大地重力测量学
geodetic height	大地高
geodetic instrument	大地测量仪器
geodetic latitude	大地纬度
geodetic leveling	大地水准测量
geodetic longitude	大地经度
geodetic network	大地网
geodetic origin	大地原点
geodetic parallel	大地纬圈
geodetic position	大地位置
geodetic reference system	大地测量参考系统
geodetic reference system 1980 (GRS1980)	1980 国际大地测量参考系统

G

geodetic surveying	大地测量,大地测量学
geodetist	大地测量学家
geodimeter	光速测距仪;光电测距仪
geoexploration	地球物理勘测
geographic center	地理中心
geographic coordinates, geographic graticule	地理坐标
geographic data	地理数据
geographic database	地理数据库
geographic elements	地理要素
geographic entity	地理实体
geographic general name	地理通名
geographic grid	地理格网
geographic identifier	地理标识符
geographic information	地理信息
geographic information communication	地理信息传输
geographic information science	地理信息科学
geographic information service	地理信息服务
geographic information system (GIS)	地理信息系统
Geographic Information/Geomatics, ISO, ISO/TC 211	国际标准化组织地理信息/地球信息技术委员会
geographic latitude	地理纬度
geographic line matching	边线匹配

G

geographic longitude	地理经度
geographic markup language (GML)	地理置标语言
geographic name	地名
geographic name database	地名数据库
geographic name index	地名索引
geographic position	地理位置
geographical azimuth	地理方位角
geographical name transcription, geographical name transliteration	地名转写
geographical parallel	地理纬圈
geographical pole	地极
geographical reference system	地理参考系
Geographical Society of China (GSC)	中国地理学会
geographical viewing distance	地理视距
geography	地理学,地理
geoid	大地水准面
geoid height, geoid undulation	大地水准面高,大地水准面差距
geoinformatics	地球空间信息科学
geological condition	地质状况
geological fault	地质断层
geological fissure	地质裂缝
geological hazard(disaster)	地质灾害
geological interpretation of photograph	像片地质判读,像片地质解译
geological map	野外地质图

geological photomap	影像地质图
geological profile survey	地质剖面测量
geological scheme	地质略图
geological section map	地质剖面图
geological survey	地质调查
geology	地质学
geomagnetic field	地磁场
geomagnetic meridian	地磁子午线
geomagnetic pole	地磁极
geomagnetic survey station	磁测站
geomatics	测绘学
geomatics engineer	测绘工程师
geomatics engineering	测绘工程
geomechanics	地质力学
GeoMedia	地理信息系统软件 GeoMedia
geometric accuracy	几何精度
geometric condition	几何条件
geometric constraint in space	空间几何约束
geometric co-registration	几何配准
geometric correction of image	影像几何纠正
geometric correction, geometric rectification	几何校正
geometric dilution of precision (GDOP)	几何精度因子
geometric distortion	几何畸变
geometric distribution	几何分布
geometric feature	几何特征

G

geometric geodesy	几何大地测量学
geometric model	几何模型
geometric modeling	几何建模
geometric optics	几何光学
geometric orientation	几何定向
geometric parameters	几何参数
geometric precise correction	几何精校正
geometric rectification of imagery	图像几何纠正
geometric registration of imagery	图像几何配准
geometric-optics model	几何光学模型
geometry	几何,几何学
geometry matching	几何匹配
geomorphological map	地貌图
geophysical effect	地球物理效应
geophysical exploration, geophysical prospecting	地球物理勘测
geophysical instrument	地球物理勘测仪
geophysical log	地球物理测井
geophysical magnetometer	地球物理磁强计
geophysical method	地球物理方法
geophysical satellite	地球物理卫星
geophysical seismic exploration	地球物理地震探测
geophysics	地球物理学
geophysist	地球物理学家
geopotential	地球位,大地位
geopotential number	地球位数
georeference	地理[坐标]参照
geo-referenced image	地学编码影像

G

geo-relational model	地理相关模型
geo-robot	测量机器人
GeoStar	吉奥地理信息管理软件
geostationary orbit (satellite)	地球静止轨道卫星
geo-synchronous satellite	地球同步卫星
geotechnical	岩土工程
geotechnical method	岩土工程方法
gimbal suspension	常平架
g-inverse	G-逆
GIPSY	美国 JPL 研发的非差精密定位与定轨软件
girder	梁,钢桁的支架
GIS	地理信息系统
glacier movement	冰川运动
glaciology	冰河学
glass filter	滤光镜
global geometrical constraint	整体几何约束
global gravity anomalies	全球重力异常
global ionosphere map(GIM)	全球电离层图
global mean sea surface	全球海平面
global navigation satellite system (GNSS)	全球导航卫星系统
global optimization	全局优化
global orbiting navigation satellite system (GLONASS)	全球轨道导航卫星系统(俄罗斯)
global plate motions	全球板块运动
global positioning system(GPS)	全球定位系统

G

global relaxation	整体松弛
global tectonic activity	整体构造活动
globe	地球仪
GLONASS	全球轨道导航卫星系统(俄罗斯)
GLONASS receiver	GLONASS 接收机
GML	地理置标语言
GMST	格林尼治平恒星时
gnomonic projection	球心投影,极平投影
GNSS	全球导航卫星系统
goal for urban redevelopment	城市发展目标
GOCE	重力卫星
golden section	黄金分割
good sky visibility	可见性好
goodness-of-fit test	拟合良好性检验,适合度检测
GPPS,WinPRISM,Solution	Ashtech GPS 接收机随机软件
GPS	全球定位系统
GPS aerotriangulation	GPS 空中三角测量
GPS carrier phase	GPS 载波相位
GPS constellation	GPS 星座
GPS receiver	GPS 接收机
GPS surveying	GPS 测量
GPS time (GPST)	GPS 时间
GPS week	GPS 周
GPST	GPS 时间
GPSurvey,TGO	Trimble GPS 接收机随机软件

graben	地堑
GRACE	重力卫星
grade	坡度,地面标高,室外地坪
grade change point	变坡点
grade location	坡度测设
gradient	梯度,坡度,斜率
gradient vector	梯度向量
gradient-based adaptive filter	基于梯度的自适应滤波
graduated scale	分画尺
gramme(g)	克
graph and image control point	图形图像控制点
graphic data	图形数据
graphic database	图形数据库
graphic display interface(GUI)	图形用户接口
graphic element	图形元素
graphic information	图形信息
graphic overlay	图形叠置
graphic sign	图形记号
graphic symbol	图形符号
graphical accuracy	图解精度
graphical adjustment	图解平差法
graphical rectification	图解纠正
graphical representation	图示,以图样表达
graphical solution	图解
graphical traversing	图解导线测量
graphics	图形

graph-paper	方格纸
GRASS	地理资源分析支持系统
grass land	草地
grating	光栅
gravimeter	重力仪
gravimetric baseline	重力基线
gravimetric control point	重力控制点
gravimetric deflection	重力偏差
gravimetric deflection of the vertical	重力垂线偏差
gravimetric geoid	重力大地水准面
gravimetric leveling	重力水准测量
gravimetric point	重力点
gravimetry	重力测量学
gravisat	重力测量卫星
gravitation	引力
gravitational acceleration	引力加速度
gravitational constant	引力常数,重力常数
gravitational field	引力场
gravitational harmonics	引力球谐函数
gravitational potential	引力位
gravitational potential function	引力位函数
gravitational torque	重力矩
gravitational vector	引力向量,重力向量
gravity	重力
gravity acceleration	重力加速度
gravity acceleration in space	空间重力加速度

gravity anomaly	重力异常
gravity anomaly vector	重力异常向量
gravity database	重力数据库
gravity datum	重力基准
gravity disturbance vector	重力扰动向量
gravity equipotential surfaces	重力扰动向量
gravity field	重力场
gravity field of the Earth	地球重力场
gravity gradient measurement, gravity gradiometry	重力梯度测量
gravity gradiometer, gradiometer	重力梯度仪
gravity measurement	重力测量
gravity network	重力[点]网
gravity observation of Earth tide	重力固体潮观测
gravity potential	重力位
gravity recovery	重力场恢复
gravity reduction	重力归算
gravity station	重力点
gravity vector	重力向量
Gray tone	灰色色调,黑白亮度等级
gray(=grey) value	灰度值
great circle	大圈,大圆
great circle distance	大圆航程
great circle sailing chart	大圆航线图
green area	绿地面积
green public space	公共绿地
greening rate	绿地率

green-planted city	绿化城市
Greenwich apparent sidereal time (GAST)	格林尼治真恒星时
Greenwich mean sidereal time (GMST)	格林尼治平恒星时
Greenwich mean time	格林尼治时间
Greenwich meridian	格林尼治子午线,起始子午线
grey correlation analyses	灰关联分析
grey correlation grade	灰关联度
grey level	灰度
grey level co-occurrence matrix (GLCM)	灰度共生矩阵
grey level morphology	灰度形态学
grey modeling	灰色模型
grey sets	灰集理论
grey system	灰色系统
grey system theory	灰色系统理论
grey value	灰度值
grey wedge, optical wedge	灰楔
grid	网格
grid azimuth	坐标方位角,格网方位角
grid bearing	坐标方位角
grid cell	网格单元
grid computing	网格计算
grid coordinate	平面直角坐标,格网坐标

G

grid map	网格地图
grid method	网格法
grid north	坐标北方向
grid of neighboring zone	邻带方里网
grid origin, origin of coordinate system	坐标原点
grid structure	网格结构
grid technology	网格技术
grid-TIN	混合形式的数字高程模型
gross area	总面积
gross building area	总建筑占地面积
gross domestic product(GDP)	国内生产总值
gross error	异常值,粗差
gross error detection	粗差检测
gross floor area	建筑总面积
gross national product(GNP)	国民生产总值
gross population density	人口总密度
gross residential area	总居住区面积
ground control point(GCP)	地面控制点
ground elevation	地面高程
ground feature	地物
ground measurement	近地表测量
ground nadir point	地底点
ground penetrating radar	探地雷达
ground receiving station	地面接收站
ground speed	地速
ground subsidence(settlement)	地面沉降

ground support system	地面服务站
ground sweeping	拖底扫海
ground tilt observation	地倾斜观测
ground truth	地面实况
ground water	地下水
ground water monitoring well	地下水观测孔
ground-based augmentation system	地基增强系统
ground-based control system (GCS)	地面控制部分
ground-based system	地基系统
groundwater condition	地下水情况
groundwater level	地下水位
group velocity	群速
grouping	自动编组
grouting	薄泥浆填塞
GRS1980	1980 国际大地测量参考系统
gruber point	标准配置点
GSC	中国地理学会
GSSP, Spectrum Survey	SOKKIA GPS 接收机随机软件
guaranteed efficiency depth datum	深度基准面保证率
GUI	图形用户接口
guide post	路标,方向标
gully shape	峡谷形态
gyro	陀螺
gyro azimuth	陀螺方位角
gyro course	陀螺罗航向

gyro mark	陀螺光标线
gyro oscillation	陀螺摆动
gyro scale	陀螺分画板
gyro spin axis	陀螺旋转轴
gyro theodolite	陀螺经纬仪
gyrocompass	陀螺罗经
gyrocompass north	陀螺北
gyroscope	陀螺仪,回转仪
gyroscopic orientation	陀螺仪定向
gyroscopic orientation survey	陀螺仪定向测量
gyroscopic theodolite	陀螺经纬仪
gyro-stabilized platform	陀螺稳定平台

H

hachuring	晕滃法
Hadamard transformation	阿达马变换
hair-pin curve location	回头曲线测设
half-interval contour	间曲线
half-mark	实测标,浮动测标
halftone	半色调
halocline	盐度跃层
hand level	手持水准仪
harbor chart	港湾图
harbor city	港口城市
harbor engineering survey	港口工程测量
harbor survey	港湾测量
hard copy	硬拷贝
harmonic analysis of tide	潮汐调和分析
hasty road	简易公路
haze detection	薄云检测
haze removal	薄云去除
HDS2000/HDS2003	中海达 GPS 数据处理软件
heading	航向
heat radiation	热辐射
heave compensation	波浪补偿
heave compensator	波浪补偿器
hectometre	百米

H

hedge	树篱，障碍物
height accuracy	高程精度
height anomaly	高程异常
height datum	高程原点
height datum, vertical datum	高程基准
height of high water	高潮高
height of instrument (HI)	仪器高
height of low water	低潮高
height of sight, elevation of sight	视线高
height of target (HT)	目标高
height system	高程系统
Helava	Leica 经销的数字摄影测量系统
Heliocentric coordinate systems	日心坐标系
Helmert'spheroid	赫尔默特椭球
hemisphere	半球体
hemisphere	半球
heptagon	七边形
heterogeneity	异质性
heterogeneous	参差的，不统一的
hexagon	六边形
hidden trouble	隐患
hiding relation	遮挡关系
hierarchical	分层
hierarchical organization	等级结构
hierarchical processing	分层处理
hierarchical structure image	分层结构影像
HIFI (the Height Interpolation by Finite Elements)	德国 Munich 大学研制的 DTM 软件

H

high density digital tape(HDDT)	高密度数字磁带
high frequency modulation	高通滤波
high latitude	高纬度
high oblique	大角度倾斜
high resolution	高分辨率
high resolution aerial image	高分辨率航空影像
high resolution cameras	高分辨率摄像机
high resolution spatial panchromatic image	高分辨率全色影像
high spatial resolution	高空间分辨率
high spectral image	高光谱图像
high speed highway	高速公路
high water	高潮
high-accurate outline	高精度轮廓
high-contrast	高对比度
higher-order triangulation	高等三角测量
high-pass filter	高通滤波器
high-pass filtering	高通滤波
high-precision leveling	高精度水准测量
high-precision location operator	高精度定位算子
high-rise building	高层建筑
high-voltage electricity	高压电
highway	公路,道路
highway bridge	公路桥
highway design	公路设计
highway network	公路网
Hilbert-Huang transform	希尔伯特-黄变换
hill shading	晕渲法

hillside 山腰,山坡
histogram 直方图
histogram equalization 直方图均衡
histogram matching 直方图匹配
histogram specification 直方图规格化
historical map 历史地图
Holder's inequality 赫耳德不等式
hollow 山谷
hologram photography 全息摄影
hologrammetry 全息摄影测量
holography 全息摄影,全息技术
homeotheric map 组合地图
homogeneity 均匀性,同性
homogeneous equation 齐次方程
homogeneous sphere 匀质球体
homologous 同名点的,同调的,类似的
homologous image points 同名像点
homologous points 同名(像)点
homologous ray 同名射线
homomorphic 同态
homomorphic filter 同态滤波
Hooke's law 虎克定律
horizon 地平面
horizon camera 地平线摄影机
horizon photograph 地平线像片
horizon-range projective imaging 平距成像
horizontal accuracy 水平精度

horizontal alignment	平面定线
horizontal angle	水平角
horizontal asymptote	水平渐近线
horizontal axis	水平轴,横轴,旋转轴
horizontal circle	水平度盘
horizontal component	水平分量
horizontal control network	平面控制网,水平控制网
horizontal control point, horizontal control station	平面控制点
horizontal coordinate, plane coordinate	平面坐标
horizontal curve layout	平面曲线测设
horizontal dilution of precision (HDOP)	水平精度因子
horizontal displacement	水平位移
horizontal displacement observation	水平位移观测
horizontal distance	平距
horizontal geodetic datum	大地测量平面基准
horizontal gradient of gravity	重力水平梯度
horizontal line	水平线
horizontal parallax	地平视差,左右视差,横视差
horizontal plane	平面
horizontal range	水平射程
horizontal refraction error	水平折光差
horizontal section	水平截面
horizontal survey	水平测量,平面测量

H

horst	地垒
hotspot effect	热点效应
hour circle	时圈
HRV(High Resolution Visible)	高分辨率可见光传感器(SPOT卫星搭载)
Huanghai mean sea level	黄海平均海[水]面
Huanghai vertical datum of 1956	1956 黄海高程系统
hue,tone	色调
human visual system(HVS)	人眼视觉特性
humidity	湿度
hybrid inversion	混合反演
hybrid point-line photogrammetry	点、线混合摄影测量
hybrid structure	混合结构
hydrant	消防栓,消防龙头
hydrographer	水道学者，水道测量家
hydrographic engineering survey	水利工程测量
Hydrographic Commission	海道测量委员会
hydrographic control point	海控点
hydrographic datum	水文测量基准面
hydrographic engineering survey	水利工程测量
hydrographic mark	海洋测量标志
hydrographic survey	海道测量,水道测量
hydrographic surveying and charting	海洋测绘
hydrographic surveying and charting database	水道测绘数据库
hydrologic features	水文要素

H

hydrologic patterns	水文特征
hydrological analysis	水文分析
hydrometry	水文观测,水文测验
hydrophone	水听器
hydrostatic leveling	液体静力水准测量
hyperbola	双曲线
hyperbolic function	双曲函数
hyperbolic navigation chart	双曲线导航图
hyperbolic positioning	双曲线定位
hyperbolic positioning system	双曲线定位系统
hyperfocal distance	超焦点距离
hypergeometric distribution	超几何分布
hypergraph	超图
hypermedia	超媒体
hyperplane segmentation	超平面分割
hyperspectral image	超谱图像,高光谱影像
hyperspectral imager	超光谱成像仪
hyperspectral imagery	高光谱
hyperspectral reflectance	高光谱反射率
hyperspectral remote sensing	高光谱遥感
hyperspectral texture	高光谱纹理
hyperspectrum	超光谱
hypocenter	(核爆炸\地震的)震源
hypocycloid	内摆线
hypotenuse	斜边
hypothesis	假设

hypothesis testing	假设检验
hypsometric layer	分层设色法
hypsometric map	地势图
hypsometric tinting	分层设色

H

I

I2S Digital Plotter	美国的卫星遥感测图处理系统
IAG	国际大地测量协会
IAU	国际天文学联合会
ICA	国际制图协会
ice-snow covered area	冰雪覆盖区
ICRF	国际天球参考架
ICSU	国际科学(联合会)理事会
idempotent	[数]幂等,等幂
idempotent matrix	幂等阵
idenfinite integration	不定积分法
identification code	识别码
identifier	标识码
identity matrix	单位矩阵
IERS	国际地球自转服务
IFOV	瞬时视场
IGS	国际 GNSS 服务组织
IGU	国际地理学联合会
IHO	国际海道测量组织
ill-posed problem	病态问题
illuminance of ground	地面照度
illuminant	发光体,光源
illuminanting engineering	照明工程

illuminated patch	光斑,光点
illuminating power	亮度
image	成像
image overlaying	图像融合
image acutance	影像清晰度
image analysis	影像分析,图像分析
image block coding	影像块编码
image blocking	影像分块
image chains (IC)	成像链
image classification	影像分类
image coding	图像编码
image compression	影像压缩
image coordinate	像片坐标,像点坐标
image correlation	影像相关
image data	图像数据
image database	影像数据库
image denoising	图像降噪
image description	图像描述
image digitization	图像数字化
image displacement	影像位移,像移
image enhancement	影像增强
image fusion	影像融合
image fusion evaluation	影像融合评价
image horizon, horizon trace, vanishing line	像地平线,合线
image information cognition	影像信息认知
image integration	影像复合
image interpolation	影像插值

I

image interpretation 影像判读
image map 影像图
image match 影像匹配
image mosaic 影像镶嵌
image motion compensation(IMC) 像移补偿
image overlaying 图像复合
image plane 像平面
image point 像点
image processing 图像处理
image pyramid 影像金字塔
image pyramid decomposition 图像金字塔分解
image quality 影像质量
image quality control 影像质量控制
image quality correction 图质改善处理
image quality measure 像质评价
image recognition 图像识别
image reconstruction 影像重建
image rectification 影像纠正
image refinement 影像精化
image registration 图像配准
image resampling by pixel doubling 双像素重采样
image resizing 图像缩放
image resolution 影像分辨率
image restoration 影像复原
image segmentation 图像分割
image sequence 序列影像
image setter 激光照排机
image shift 像移

image simulation	图像模拟
image space	像空间
image space coordinate system	像空间坐标系
image spatial resolution	图像空间分辨率
image strip	图像条带
image transformation	图像变换
image transmission	图像传输
image understanding	图像理解
image, imagery	影像,图像
imager	成像仪
ImageStation	美国/德国 Z/I Imaging 公司的卫星遥感测图处理系统
imaginary number	虚数
imaginary part	虚部
imaginary root	虚根
imaging equation	构像方程
imaging Fourier transform spectrometer	成像干涉光谱仪
imaging model	成像模型
imaging radar	成像雷达
imaging spectrometer	成像光谱仪
imaging system	成像系统
immediate plan	近期建设规划
immune algorithms	免疫算法
IMO	国际海事组织
impedance	阻碍强度
implicit function	隐函数

I

imposed load	外加荷载,附加荷载
impossible event	不可能事件
improper fraction	假分数
improper integral	广义积分
improvement area	改建地区
improvement of satellite orbit	卫星轨道改进
impulse noise	椒盐噪声
impulse response fuction	脉冲响应函数
IMU	惯性测量装置
incidence	入射
incident angle	入射角
inclination angle, tilt angle	倾角
inclination, tilt	倾斜,斜角
inclined plane	斜面
inclined shaft	斜井
inclinometer and tiltmeter	测斜仪
inconsistency	不一致性
increasing function	递增函数
increasing series	递增级数
increment	增量
indefinite integral	不定积分
indentification	判读,识别
independent baseline	独立基线
independent component analysis (ICA)	独立分量分析
independent coordinate system	独立坐标系
independent model aerial triangulation	独立模型法空中三角测量

independent random variable	独立随机变量
independent triangulation	独立三角测量
indeterminacy	不确定，不明确
index contour	计曲线,加粗等高线
index diagram, sheet index	图幅接合表
index error of vertical circle	竖盘指标差
index mosaic	镶嵌索引图
index of precision	精度指标
indicatrix	变形椭圆
indirect illumination	间接照明
indirect scheme of digital rectification	间接法纠正
indirection adjustment	间接平差
individual element classification	单要素分类
indoor control field	室内控制场
indoor test field	室内实验场
Indu SURF (Industrial Surface Measurement)	Zeiss 的解析测图仪 C100 附加一对 CCD 相机构成的混合数字摄影测量系统
induced polarization(IP)	激发极化
industrial architecture	工业建筑
industrial city	工业城市
industrial district	工业区
industrial land	工业用地
industrial measuring system	工业测量系统
industrial photogrammetry	工业摄影测量
industrial sewage	工业污水

I

industrial sudden accidents	工业突发事故
inequality	不等式
inequality sign	不等号
inertia	惯性
inertial coordinate system	惯性坐标系
inertial navigation system(INS)	惯性导航系统
inertial reference system	惯性参考系统
inertial reference unit(IRU)	惯性参考设备
inertial space	惯性空间
inertial surveying system(ISS)	惯性测量系统
inference decision	推理判决
infill system	填充体系
inflection point	拐点
information acquisition	信息获取
information age	信息时代
information attribute	信息属性
information compound	信息复合
information disturbance accident	信息干扰事件
information entropy	信息熵
information extraction	信息提取
information fusion	信息融合
information grid	信息网格
information hiding	信息隐藏
information safty	信息安全
information sharing	信息共享
infrared cloud imagery	红外卫星云图
infrared EDM instrument ,infrared ranger	红外测距仪

infrared film	红外片
infrared imagery	红外图像
infrared photography	红外摄影
infrared radiation	红外辐射
infrared radiometer	红外辐射计
infrared remote sensing	红外遥感
infrared scanner	红外扫描仪
infrared spectrum	红外波谱
infrared window	红外窗口
inherent error	固有误差
initial approximation	初始近似值
initial azimuth	起始方位角
initial condition	原始条件,初始条件
initial origin	始点,起点
initial value	初值,始值
initial velocity	初速度
initialization	初始化
initial-value problem	初值问题
inner city	城市中心区
inner constraints	内约束
inner core	内核
inner product	内积
in-place construction	现场施工
input box	输入框
input signal	输入信号
InSAR	干涉合成孔径雷达,雷达干涉测量
inscribed circle	内切圆

inscribed triangle	内接三角形
insignificant	不显著的
insolation standard	日照标准
inspection hole	检修孔
inspection manhole	检修井
instability	不稳定性
installation survey	安装测量
instantaneous acceleration	瞬时加速度
instantaneous field of view(IFOV)	瞬时视场
instantaneous map	瞬间地图
instantaneous pole	瞬时极
instantaneous sea level	瞬时海面
instantaneous speed	瞬时速率
instantaneous velocity	瞬时速度
instrument adjustment	仪器校正
instrument height	仪器高
instrument of geomatics engineering	测绘仪器
instrument testing	仪器检测
instrumental constant	仪器常数
instrumental error	仪器误差
insular shelf	岛架
integer	整数
integer ambiguity resolution	整周模糊度解算
integer part	整数部分
integer solution	整数解
integer value	整数值
integer wavelet transform	整数小波变换
integrable function	可积函数

integrant	被积函数
integrated data	集成数据
integrated geodesy	整体大地测量
integrated monitoring method	综合监测方法
integrated navigation	组合导航
integrated navigation system	组合导航系统
integrated positioning	组合定位
integrated positioning systems	集成定位系统
integrity	完好性
intelligent agent	智能主体
intelligent photogrammetry	智能摄影测量
intelligent simulation of structure state	结构状态智能模拟
intelligent transportation system (ITS)	智能交通系统
intelligent visualization	智能可视化
intended course	计划航向
intensity	强度
intensity image	反射强度像
intensity value	强度值,亮度值
interactive	交互式的
interactive processing	人机交互处理
intercept	截距
inter-city transportation	城际交通枢纽
intercom	内部通信联络系统,联络用对讲电话装置
interface option	界面接口选项
interference sidescan sonar	干涉侧扫声呐

I

interferogram	干涉图
interferogram fringe	相干条纹
interferometric fringe	干涉条纹
interferometric synthetic aperture radar(InSAR)	干涉合成孔径雷达，雷达干涉测量
interferometry	干涉技术
interior angle	内角
interior orientation	内部定向
inter-island sea	岛间海
interlaced sampling	交错采样
interline of sounding	加密测深线
interline transfer	行间传输
intermediate contour	首曲线，基本等高线
internal accuracy	内部精度
internal energy	内能
internal force	内力
internal memory	内存
internal orientation	内定向
internal reliability	内部可靠性
internal wave	内波
International Association for the Physical Science of the Ocean	国际海洋物理科学协会
International Association of Geodesy(IAG)	国际大地测量协会
International Astronomical Union (IAU)	国际天文学联合会
international atomic time(TAI)	国际原子时
international bathymetric chart	国际海底地形图

International Cartographic Association(ICA)	国际制图协会
international celestial reference frame(ICRF)	国际天球参考架
international chart	国际海图
International Council for Exploration of the Sea	国际海洋考察理事会
International Council for Science, International Council of Scientific Unions(ICSU)	国际科学(联合会)理事会
International Earth Rotation and Reference Service (IERS)	国际地球自转服务
International Federation of Surveyors(FIG)	国际测量师联合会
International Geographical Union (IGU)	国际地理学联合会
international GNSS service (IGS)	国际 GNSS 服务组织
International Hydrographic Bureau	国际海道测量局
International Hydrographic Organization(IHO)	国际海道测量组织
International Maritime Organization (IMO)	国际海事组织
International Organization for Standardization(ISO)	国际标准化组织
International Radio Maritime committee	国际海上无线电通讯委员会
International Sea Bed Authority	国际海床管理局

I

International Society for Photogrammetry and Remote Sensing (ISPRS)	国际摄影测量与遥感学会
international spheroid	国际椭球
international terrestrial reference frame(ITRF)	国际地球参考框架
international terrestrial reference system (ITRS)	国际地球参考系
International Union of Geodesy and Geophysics(IUGG)	国际大地测量与地球物理联合会
international waters	公海
interometry SAR	干涉雷达
interoperability	互操作
interpolating polynomial	插值多项式
interpolation	插值
interpolation base function	插值基函数
interpolation point	内插点
interpretation	判读,判释,解释
interpretation element	判读要素
interpretation of echograms	声图判读
interpretoscope	判读仪
interprovincial highway	省际公路
interpupilliary adjustment	眼基线调节
interrupted projection	分瓣投影
interscale model	尺度间相关模型
intersection	交会法
intersection	前方交会
intersection angle	交会角

intersection angle of LOP 位置(线交)角
intertidal zone 潮间带
interurban railroad 城际铁路
interval 区间
interval estimate, interval estimation 区间估计
interval of survey line 测深线间隔
interval scaling 等距量表
intracity traffic 市内交通
intrascale model 尺度内相关模型
invar barcode rod 铟瓦条形码尺
invar baseline wire 铟瓦基线尺
invar rod 铟瓦尺
invariable moment of histogram 直方图不变矩
invariance 不变性
invariant moment 不变矩
inverse 逆的
inverse circular function 反三角函数
inverse cosine function 反余弦函数
inverse Fourier transformation 傅立叶逆变换
inverse function 反函数
inverse matrix 逆矩阵
inverse of a matrix 矩阵的逆
inverse of weight matrix 权逆阵
inverse photogrammetry 逆反摄影测量
inverse plummet observation 倒锤[线]观测,倒锤法
inverse problem 逆算问题

inverse relation 逆关系
inverse route 航线反转
inverse solution of geodetic problem 大地主题反解
inversely proportional 反比例
inverse-sine function 反正弦函数
inverse-tangent function 反正切函数
inversion of component temperature 组分温度反演
inverted pendulum 倒锤[线]观测,倒锤法
invertibility 可逆性
invertibility condition 可逆性条件
invertible 可逆的
ion concentration 离子浓度
ionosphere 电离层
ionosphere-free combination 无电离层组合
ionospheric delay 电离层延迟
ionospheric refraction 电离层折射
ionospheric refraction correction 电离层折射改正
IR (heat) radiation 红外(热)辐射
IRI-D256 加拿大的实时摄影测量系统
irrational equation 无理方程
irrational number 无理数
irreducibility 不可约性
irregular 不规则
irregular points 不规则离散点
island 岛屿

island arc　岛弧
island chain　岛链
island chart　岛屿图
island survey　岛屿测量
island-mainland connection survey　岛屿联测
ISO　国际标准化组织
isobath　等深线
isocenter of photograph　像等角点
isodynamic line　等磁力线
isolated network　独立网
isolated node　孤立节点
isoline　等值线
isoline map　等值线地图
isoline method　等值线法
isometric latitude　等量纬度
isometric parallel　等比线
isometry　等高
isomorphism　同构
isosalinity line　等盐线
isosceles triangle　等腰三角形
isostasy　地壳均衡
isostatic correction　地壳均衡改正
isostatic geoid　均衡大地水准面
isotherm　等温线
ISPRS　国际摄影测量与遥感学会
ISS　惯性测量系统
item under construction　在建项目

iteration	迭代
iteration method with variable weights	选权迭代法
iterative (trial and error) approach	模拟(迭代)法
iterative algorithm	迭代算法
iterative closest point algorithm (ICP)	最近邻迭代
iterative inversion	迭代反演
iterative method	迭代法
iterative projection	迭代投影
iterative self organizing data analysis techniques algorithm (ISODATA)	迭代自组织数据分析算法
iterative unmixing	迭代分解
ITRF	国际地球参考框架
ITRS	国际地球参考系
ITS	智能交通系统
IUGG	国际大地测量与地球物理联合会

J

JARS	日本发射的地球资源卫星
Javad	伽瓦德
JERS-1	日本地球资源卫星合成孔径雷达
jiont distribution	联合分布
jiont distribution function	联合分布函数
junction	交叉口
junction point of traverse	导线节点

K

KARS	UNB 大学 GPS 数据处理软件
Kelvin	绝对温度单位
Kepler's Laws of Planetary Motion	开普勒行星运动定律
key city	中心城市
key frames	关键帧
key function	主要功能
key laboratory	重点实验室
key word	关键字
kilogramme	公斤，千克
kilometer	公里，千米
kilometer grid	方里网
kilometer stone	里程碑
kinematic	动态的
kinematic positioning	动态定位
kinetic	动力(学)的
kinetic equilibrium	动力平衡，动态平衡
Kingspad	加拿大卡尔加里大学研制 GPS 数据处理软件
K-means algorithm	K-均值算法
knoll	海丘
knowledge reasoning	知识推理
knowledge representation	知识表示

known station	已知站(点)
Krasovsky ellipsoid	克拉索夫斯基椭球
Kriging	克里金法
kurtosis	峰度

K

L

L1 frequency	GPS 信号频率之一(1575.42 MHz)
L2 frequency	GPS 信号频率之一(1227.6 MHz)
LAAS	局域增强系统
labour intensive	费力
Lacoste-Romberg marine gravimeter	L&R 海洋重力仪
lacoste-Rormberg gravimeter	拉科斯特-隆贝格重力仪
lag	迟延，滞后
Lagrange interpolating polynomial	拉格朗日插值多项代
Lagrange theorem	拉格朗日定理
lake sediment	湖泊沉积
Lambert projection	兰勃特投影
lamp post	路灯
land surface water index	陆表水指数
land acquisition	土地征用
land appraisal	土地估价,土地评估
land boundary	地界
land boundary map	地类界图
land boundary survey	地界测量
land classification	土地分类
land conservation	土地保护

L

land cost	地价
land cover	土地覆盖
land degradation	土地退化
land division	土地分割
land evaluation	土地估价,土地评估
land feature	地物,地貌
land for building	建筑用地
land information system(LIS)	土地信息系统
land investigation	土地调查
land leasing	土地出让
land management	土地管理
land ownership	土地所有权
land planning survey	土地规划测量
land register	地籍簿
land resources	土地资源
land surface temperature	陆地表面温度
land surface temperature retrieval	地表温度反演
land survey	土地测量
land transfer	土地转让
land use analysis	土地利用分析
land use monitoring	土地利用动态监测
land use permit	建设用地规划许可证
land use planning	土地利用规划
land use survey	土地利用调查
landform	地形,地貌,陆地轮廓
land-ground measurement data	地表实测数据
land-leveling operation	土地平整
landsat	陆地卫星(美国)

landscape	景观
landscape pattern	景观特征
landslide	滑坡
landslide monitoring	滑坡监测
Laplace azimuth	拉普拉斯方位角
Laplace point	拉普拉斯点
Laplace surface harmonics	拉普拉斯面调和函数,拉普拉斯面谐函数
large footprint lidar	大光斑激光雷达
large format camera(LFC)	大像幅摄影机
large scale	大比例尺
large scale topographical map	大比例尺地形图
large-sample	大样本
laser alignment	激光准直
laser alignment equipment,laser aligner	激光准直仪
laser altimeter	激光测高仪
laser altimetry	激光测高
laser bathymetry	激光测深
laser beam	激光束
laser distance measuring instrument,laser ranger	激光测距仪
laser dot	激光点
laser eyepiece	激光目镜
laser guide of vertical shaft	竖井激光指向[法]
laser gyroscope	激光陀螺仪
laser image radar	激光成像雷达

laser level	激光水准仪
laser orbit determination	激光定轨
laser plotter	激光绘图机
laser plumbing	激光投点
laser plummet	激光铅直仪,激光对中仪
laser pulse rate	激光脉冲频率
laser rangefinder, laser ranger, laser distance measuring instrument	激光测距仪
laser ranging	激光测距
laser scanner	激光扫描仪
laser scanning	激光扫描
laser sounder	激光测深仪
laser theodolite	激光经纬仪
laser transmitter	激光发射器
LAST	地方真恒星时
last echo(return)	尾次回波
latent root	本征根,特征根
lateral error of traverse	导线横向误差
lateral overlap	旁向重叠
lateral tilt	旁向倾角,横摇
latitude	纬度
latitude circle	纬圈,平行圈
latitude of pedal	底点纬度
law of large number	大数定理
law of probability	概率论
law of universal gravitation	万有引力定律
layer depth chart	跃变层深度图

layered extraction 分层提取
layout 定线,放样
layout of an angle 角度测设
layout of curves 曲线测设
layout plan 平面布置图
layout table 放样表
layout technique 放样技术
L-band L 波段
LBS 基于位置服务
LCD 液晶显示器
lead 水砣
lead sounding 水砣测深
lead line 测深索
leading beacon 导标
leading line 导航线,叠标线
learning algorithm 学习算法
least squares collocation 最小二乘配置法,最小二乘拟合推估法
least squares correlation 最小二乘相关
least squares estimation 最小二乘估计
least squares image matching 最小二乘影像匹配
least squares matching 最小二乘匹配
least squares method(LSM) 最小二乘法
least squares spectral analysis 最小二乘谱分析
least squares spectrum 最小二乘谱
least squares template matching 最小二乘模板匹配
least-cost path 最少成本路径
LED 发光二极管

left femur	左侧
left-hand limit	左方极限
left-handed coordinate system	左手坐标系
left-hand-side(L. H. S)	左手边
legislation on urban planning	城市规划法规
Leica	瑞士徕卡
lengend	图例
length	长(度)
length of curve	曲线长度
lens distortion	镜头畸变
lens shutter	中心式快门
LEO	低轨
LEOS	低轨卫星
level	水准仪
level of detail	层次结构
level sensor	水平传感器
level spheroid	水准椭球
level surface	水准面
level terrain	平坦地
leveler	水准测量员
leveling	水准测量
leveling control	水准测量控制
leveling line	水准路线
leveling network	水准网
leveling of instrument	仪器置平
leveling origin	水准原点
leveling rod, leveling staff	水准尺
leveling the instrument	整平仪器

LFC	大像幅摄影机
LGO	Leica GPS 接收机随机软件
licensed engineer	注册工程师
LIDAR	激光探测与测距(激光雷达)
lifting capacity	承重能力
lifting scheme	提升算法
lifting scheme wavelet	提升小波
light beacon	灯标
light buoy	灯浮标
light characteristic	灯质
light color	灯色
light detection and ranging (LIDAR)	激光探测与测距(激光雷达)
light period	灯光周期
light rail transit	轻轨交通
light range	灯光射程
light ship	灯船
light tone	明亮色调,浅色调
light-emitting diode(LED)	发光二极管
lighthouse	灯塔
lightness	亮度
likehood function	似然函数
limb scanning	临边扫描
limit	极限
limit error	极限误差
limit theorems	极限定理

limiting danger line	危险界限
limiting friction	最大静摩擦
line coverage	线图层
line diffusion function model	线扩散函数
line hydrophone	线列水听器
line jitter	行抖动差
line map	线画地图
line match	线段匹配
line moment	线矩
line of best-fit	最佳拟合
line of intersection	交线
line of position（LOP）	位置线
line perspective	线中心投影
line photogrammetry	直线摄影测量
line scanning	线扫描
line segment	线段
line smoothing	曲线光滑
line stereo matching	直线相关
line symbol	线状符号
line target	线状目标
linear array	线阵列
linear array push-broom imagery, linear pushbroom imagery	线阵推扫式影像
linear array sensor, push-broom sensor	线阵遥感器
linear CCD push-broom	线阵 CCD 推扫式
linear convergence	线性收敛性
linear correlation	线性相关

L

linear differeantial equation	线性微分方程
linear equation	线性方程
linear equation in two unknowns	二元一次方程
linear error of traverse	导线长度误差
linear filter	线性滤波器
linear filter model	线性滤波模型
linear frequency modulation signal	线性调频信号
linear intersection	边交会法
linear models	线性模型
linear multiscale transform	线性多尺度变换
linear nonstationary process	线性非平稳过程
linear process	线性过程
linear regression	线性回归
linear spectral unmixing	线性光谱混合求解方法
linear stationary model	线性平稳模型
linear triangulation chain	线形锁
linear triangulation network	线形网
linear unbiased estimator	线性无偏估计
linear-angular intersection	边角交会法
linearized math model	线性化的数学模型
linearly dependent	线性相关的
linearly independent	线性无关的
line-like object	线状地物
lining	衬砌
LIP	原武汉测绘科技大学GPS数据处理软件
liquid crystal display (LCD)	液晶显示器

LIS	土地信息系统
list	横倾
list of radio beacon	无线电指向标表
lithospheric structure	岩石圈结构
litre	升
littoral topographic survey	海岸带地形测量
littoral zone topographic chart	海岸带地形图
living density	居住密度
living floor space	居住使用面积
living water use	生活用水
LLR	激光测月
lmage interpolation	图像插值
LMST	地方平恒星时
load	负荷
load bearing	承重,承载能力
load bearing wall	承重墙
lobe	波瓣
local apparent sidereal time (LAST)	地方真恒星时
local apparent time	地方视时
local area augmentation system (LAAS)	局域增强系统
local astronomical time	地方天文时
local automatic searching	局部自动搜索
local city	地方城市
local datum	地方基准,局部基准
local differential GPS	局域差分 GPS
local enhancement and filtering	局部增强滤波

local matching	局部匹配
local maximum	局部极大(值)
local mean sea level	当地平均海面
local mean sidereal time(LMST)	地方平恒星时
local meridian	地方子午线
local minimum	局部极小(值)
local navigation satellite system	区域导航卫星系统
local planning	局部规划
local precision estimate	局部精度估计
local sidereal time	地方恒星时
local solar time	地方太阳时
local standard time	地方标准时
local time	地方时
local true time	地方真时
locating of noise pixel	噪声定位
location	地点,位置
location map	位置图
location of pier	桥墩定位
location of route	线路中线测量
location-based services(LBS)	基于位置服务
locus	轨迹
Lodriguez matrix	罗德里格矩阵
logarithm	对数
logarithmic distribution	对数分布
logarithmic equation	对数方程
logarithmic function	对数函数
logical consistency	逻辑兼容,逻辑一致性

logical deduction 逻辑推论
logistic model 逻辑斯蒂法
long baseline acoustic system 长基线水声定位系统
long division method 长除法
long level bubble 照准部水准器,长水准器
Long Range Navigation (LORAN) 罗兰导航系统
longitude 经度
longitude circle 经度圈
longitude of ascending node 升交点黄经
longitudinal error of traverse 导线纵向误差
longitudinal gradient 纵坡
longitudinal profile 纵断面图
longitudinal section 纵截面
longitudinal tilt, pitch 航向倾角
longitudinal view 纵视图
long-period free oscillation 长周期自由振荡
long-range EDM instrument 远程电子测距仪
long-range navigation system 长距离导航系统
long-range positioning system 远程定位系统
LOP 位置线
LORAN 罗兰导航系统
Loran chart 罗兰海图
Loran-C positioning system 罗兰-C 定位系统
loss of lock 失锁
lossless compression 无损压缩
lossy compression 有损压缩
lot 场地,工地

low altitude	低空
low coastline	低岸线
low earth orbit (LEO)	低轨
low earth orbit satellite (LEOS)	低轨卫星
low labour intensity	劳动强度低
low latitude	低纬度
low obligue	小角度倾斜
low rise building	低层建筑
low water	低潮,低水位
low water datum	低水位基准面
low water line	低潮线
lower high water	低高潮
lower layer	底层
lower limit	下限
lower low water	低低潮
lower transit	下中天
lower triangular matrix	下三角阵
lowest low water	最低低潮面
low-pass filter	低通滤波器
low-pass filtering	低通滤波
LSM	最小二乘法
lunar exploration	探月
lunar geodesy, selenodesy	月面测量学
lunar laser ranging(LLR)	激光测月
lunar latitude	月球纬度
lunar longitude	月球投影经度
lunar orbiter	月球轨道飞行器
lunar parallax	月球视差

L

lunar spatial probes	探月航天器
lunar tide	太阴潮
lunar time	太阴时
lunar topography surveying	月球地形测绘
lunisolar gravitational perturbation	日月引力摄动
lunisolar perturbation	日月摄动
lunisolar precession	日月岁差

M

macrophotogrammetry	超近景摄影测量
Magellan	麦哲伦(美国)
magnetic anomaly	磁力异常
magnetic anomaly area	磁力异常区
magnetic anomaly profile	磁异常剖面图
magnetic azimuth	磁方位角
magnetic bearing	磁象限角
magnetic compass	磁罗经
magnetic course	磁航向
magnetic declination variation	短期磁偏差
magnetic declination	磁偏角
magnetic element gradient	地磁要素梯度
magnetic elements	地磁要素
magnetic field strength	磁场强度
magnetic gradiometer	磁力梯度仪
magnetic inclination	磁倾角
magnetic latitude	磁纬度
magnetic line of force	磁力线
magnetic meridian	磁子午线
magnetic north	磁北
magnetic orientation	磁定向
magnetic pole	磁极
magnetic secular change	长期磁变
magnetic sounder	磁测深仪

magnetic station	地磁台
magnetic variation chart	磁差图
magnetized needle	磁针
magnetograph	地磁记录仪
magnetometer	磁力仪,地磁仪
magnetometry	地磁测量
mainstation	主台
main axis	主轴
main diagonal	主对角线
main check comparison	主检比对
major project	重点工程,大型项目
major street	主要街道
manhole	人孔, 检修孔
man-made activities	人类活动
man-made environment	人工环境
man-made hazards	人为灾害
man-made objects	人工地物
man-made structure	人工建筑物
mantle	地幔
manual adjustment	手调,人工调整
manual reading	人工读数
map	地图
map accuracy	地图精度
map analysis	地图分析
map appearance	地图整饰
map border	图廓
map catalog	地图目录
map clarity	地图清晰性

map collection 地图收集
map color atlas 地图色谱
map color standard 地图色标
map compilation 地图编绘
map complexity 地图复杂性
map cover 图廓覆盖
map coverage 地图图层
map data structure 地图数据结构
map database management system 地图数据库管理系统
map design 地图设计
map digitizing 地图数字化
map display 地图显示
map distortion 地图变形
map editing 地图编辑
map editorial policy 地图编辑大纲
map generalization 地图简化
map grid 地图格网
map interpretation 地图判读
map join 地图拼接
map legend 图例
map legibility 地图易读性
map library 地图库
map load 地图负载量
map making 地图制图
map manuscript 地图原图（包括实测、编绘、清绘原图）
map origin 地图坐标原点
map overlay analysis 地图叠置分析

map perception	地图感受
map plotting	地图绘制,绘图
map printing	地图印刷
map projection	地图投影
map projection classfication	地图投影分类
map projection distortion	地图投影变形
map query	地图查询
map reading	地图阅读
map reproduction	地图复制
map revision	地图更新
map scale	地图比例尺
map specification	地图规格
map symbol base	地图符号库
map symbols	地图符号
map title	图名
map use	地图利用
MapGIS	中地地理信息系统软件
MapInfo	地理信息系统软件 MapInfo
mapping	测图,制图
Mapsat	测图卫星
mapsheet	图幅
marginal sea	边缘海
marine acoustic techniques	海洋声学技术
marine acoustics	海洋声学
marine airborne remote sensing	海洋航空遥感
marine atlas	海图集

M

marine biological chart	海洋生物图
marine bottom proton sampler	海洋质子采样器
marine cesium magnetometer	海洋铯光泵磁力仪
marine charting	海洋测绘
marine charting database	海洋测绘数据库
marine climate	海洋气候
marine demarcation survey	海洋划界测量
marine earthquake chart	海洋地震图
marine ecology	海洋生态学
marine electromagnetics	海洋电磁学
marine engineering survey	海洋工程测量
marine environmental chart	海洋环境图
marine geodesy	海洋大地测量
marine geodetic control network	海洋大地测量控制网
marine geography	海洋地理学
marine geological survey	海洋地质调查
marine geology	海洋地质学
marine geophysical chart	海洋地球物理图
marine gravimeter	海洋重力仪
marine gravimetry	海洋重力测量
marine gravimetry traverse line design	海洋重力测线布设
marine gravity	海洋重力
marine gravity anomaly	海洋重力异常
marine gravity anomaly chart	海洋重力异常图
marine gravity base point	海洋重力测量基点
marine hydrological chart	海洋水文图
marine leveling	海洋水准测量

marine magnetic anomaly	海洋磁力异常
marine magnetic anomaly chart	海洋磁力异常图
marine magnetic chart	海洋地磁图
marine magnetic survey	海洋磁力测量
marine magnetometer sweeping	海洋磁力仪扫海
marine meteorological chart	海洋气象图
marine meteorology	海洋气象学
marine navigation	海上导航
marine optical instrument	海洋光学仪器
marine proton magnetometer	海洋质子磁力仪
marine resource chart	海洋资源图
marine survey	海洋测量
marine survey positioning	海洋测量定位
marine thematic survey	海洋专题测量
Markov chain	马尔柯夫链
Markov process	马尔柯夫过程
Markov random field(MRF)	马尔柯夫随机场
mask	蒙片
masking technique	掩膜技术
massive data set	海量数据
master compass	主罗经
master control station	主控站
master plan	总平面图，总体规划
master planning outline	城市总体规划纲要
match window	匹配窗口
matched filter	匹配滤波
material storage	物资储备
mathematical expectation	数学期望

mathmatics 数学
matrix 矩阵
matrix addition 矩阵加法
matrix equation 矩阵方程
matrix multiplication 矩阵乘法
matrix operation 矩阵运算
maximum 极大
maximum absolute error 最大绝对误差
maximum a-posteriori probability 最大后验概率
maximum entropy 最大熵
maximum likelihood estimation 最大似然估计
maximum likelihood classification 最大似然分类
max-spectrum image registration 最大谱图像配准
M-band wavelet 多进制小波
mean azimuth 平均方位角
mean earth orbit satellite 中轨道卫星
mean gravity 平均重力
mean half-tide level 平均半潮面
mean high water 平均高潮
mean high water interval 平均高潮间隙
mean high water neaps 小潮平均高潮位
mean high water springs 大潮平均高潮位
mean higher high water 平均高高潮
mean longitude 平经度,平黄经
mean low water 平均低潮
mean low water interval 平均低潮间隙
mean low water neaps 小潮平均低潮位
mean low water springs 平均大潮低潮面

mean lower low water	平均低低潮面
mean lower low water springs	平均大潮低低潮面
mean meridian	平子午线
mean neap range	平均小潮差
mean normal gravity	平均正常重力
mean parallax	平均视差
mean pole	平极
mean pole of epoch	历元平极
mean pole of rotational axis	自转轴平极
mean profile curvature	平均剖面曲率
mean radius of curvature	平均曲率半径
mean sea level(MSL)	平均海水面
mean sea surface	平均海面
mean shift model	均值漂移模型
mean sidereal time	平恒星时
mean solar day	平太阳日
mean solar time	平太阳时
mean spring range	平均大潮差
mean square error of a point	点位中误差
mean square error of angle observation	测角中误差
mean square error of azimuth	方位角中误差
mean square error of coordinate	坐标中误差
mean square error of height	高程中误差
mean square error of length	边长中误差
mean square error(MSE)	中误差
mean tide level	平均潮汐面
mean time	平时

M

mean value	均值
mean value theorem	中值定理
mean vernal equinox	平春分点
mean water level	平均水面
measurement	测量
measurement accuracy, observation accuracy	观测精度
measurement frequency	测量频率
measurement of vibrations of tall buildings	高层建筑物振动测量
measuring bar	测杆
measuring mark	测标
measuring range	量程范围
mechanisms and physics of the deformation	变形机理
median filter	中值滤波
medium Earth orbit (MEO)	中地球轨道
medium scale	中比例尺
medium-range positioning system	中程定位系统
membership	隶属度
membership function	隶属函数
mental map	心象地图
MEO	中地球轨道
Mercator projection	墨卡托投影
meridian	子午圈
meridian arc	子午线弧
meridian convergence	子午线收敛角
meridian convergence, grid convergence	子午线收敛角,坐标纵线偏角

meridian curvature	子午线曲率
meridian direction	子午方向
meridian plane, meridian section	子午面
meridian radius of curvature	子午圈曲率半径
meridian transit, transit	中天
meridional height	子午高度角(天体中天的高度角)
mesh simplification	格网简化
metadata	元数据
metadata element	元数据元素
metadata entity	元数据实体
metadata for geographic information	地理信息元数据
metadata schema	元数据模式
metadata section	元数据子集
metamerism	同色异谱
meteorological buoy	气象浮标
meteorological radar	气象雷达
meteorological satellite	气象卫星
meter	米
method by hour angle of Polaris	北极星任意时角法
method by series, method of direction observation	方向观测法
method in all combinations	全组合测角法
method of deflection angle	偏角法
method of direction observation	方向观测法
method of interpolation	插值法
method of moving target	活动觇牌法
method of parameter eatimations	测量误差理论

M

method of substitution	代换法,换元法
method of superposition	迭合法
method of tension wire alignment	引张线法
metric calibration	测量校正
metric camera	量测摄影机
metric model	度量模型
metric relation	度量关系
metric space	度量空间
microcopying, microphotography	缩微摄影
microdensitometer	测微密度计
microfilm map	缩微地图
microgravimetry	微重力测量
micrometer	读数显微器
microphotogrammetry	显微摄影测量
microscopic photograph	微观照片
MicroStation	Intergraph 公司推出的微机版三维计算机辅助设计系统
microwave distance measuring instrument	微波测距仪
microwave imagery	微波图像
microwave polarization ratio	微波极化比
microwave radiation	微波辐射
microwave radiometer	微波辐射计
microwave rangefinder	微波测距仪
microwave remote sensing	微波遥感
microwave remote sensor	微波遥感器
microwave scanner	微波扫描仪

middle tone	中性色调,灰色调
mid-oceanic ridge	大洋中脊
midpoint of curve	曲中点
military chart	军用海图
military map	军用地图
milligramme(mg)	毫克
milliliter(ml)	毫升
millimeter(mm)	毫米
mine survey	矿山测量
mineral deposits geometry	矿体几何[学]
mineral deposits map	矿产图
mineral extraction	矿物开采
minimum	极小
minimum bounding rectangle	最小限定矩形
minimum constraint adjustment	最小约束平差
minimum constraints	最小约束
minimum distance classification	最小距离分类
minimum mapping unit	最小制图单元
minimum value	极小值
minimum variance unbiased estimation	最小方差无偏估计
mining area control survey	矿区控制测量
mining map	矿山测量图
mining subsidence map	开采沉陷图
mining subsidence observation	开采沉陷观测
mining surveying	矿山测量学
mining yard plan	矿场平面图
minor angle method	小角度法

minor axis	短轴
mirror reflection	镜面反射
mirror reverse, mirror image	反像
misalignment	安置不准
misclosure	闭合差
misclosure in azimuth	方位角闭合差
misclosure in leveling	水准闭合差
misclosure of angles	角度闭合差
misclosure of horizontal angles	水平角闭合差
misclosure of leveling loop	水准环闭合差
misclosure of traverse	导线闭合差
mixed classification method	复合分类
mixed image cells, mixed pixel	混合像元
mixed spectrum	混合光谱
mixed tide harbor	混合潮港
mm-wave	毫米波
mobile mapping vehicle	测绘车
model base	模型底图
model building	建模
model combination	模型融合
model point	模型点，空间点
model space	模型空间
modeling	建模
modern civilization	现代文明
modulation	调制
modulation frequency	调频频率
modulation transfer function(MTF)	调制传递函数
modulator	调制器

moiré fringe	波纹条纹
moiré topography	莫尔条纹图,叠栅条纹图
moisture	湿度
moisture index	湿度指数
moment	矩
moment matching	矩匹配
moment of inertia	惯性矩
momentum	动量
momentum of inertial	转动惯量
monitor	监测,监视,监视器,监测器
monitor station	监控站
monitoring equipment	监测仪器
monitoring markers	监测标志
monitoring network	监测网
monitoring station	监测站
monitoring technique	监测技术
monochrome	单色
monocular image computer vision	单像计算机视觉
monomial	单项式
monotone	单调
monotonic convergence	单调收敛性
monotonic function	单调函数
monotonicaly decreasing	单调递减
monotonicaly decreasing function	单调递减函数
monotonicaly increasing	单调递增
monotonicaly increasing function	单调递增函数

Monte-carlo method	蒙特-卡罗方法
monthly mean sea level	月平均海面
moon's transit	月球中天
Moore-Penrose inverse	Moore-Penrose 广义逆
morphological gradient	形态梯度
morphological wavelet	形态小波变换
morphology	形态学
morphology transformation	形态变换
morphometric map	地貌形态示量图
mosaic	镶嵌
mosaic assembling	镶嵌处理
most probable value(MPV)	最或然值
moteorogical observation	气象观测
motion compensation	运动补偿
motion detection	运动检测
motion estimate	运动模型估计
motion parallax / pseudo parallax	移位视差 / 伪视差
moving average method	移动拟合法
moving average model	滑动平均模型
moving average parameter	滑动平均参数
moving object	运动对象
moving target detection	运动目标检测
moving-coil transducer	动圈换能器
MPV	最或然值
MRF	马尔柯夫随机场
MSE	中误差
MSL	平均海水面
MSS	多谱段扫描仪

mud snapper 采泥器
multi band polarimetric scatterometer 多波段极化散射计
multi-angle imaging spectroraiometer(MISR) 多角度光谱成像仪
multi-antenna GPS system 多天线 GPS 系统
multi-band remote sensing 多波段遥感
multibeam bathymetric system 多波束测深系统
multibeam echosounding 多波束测探
multibeam sonar 多波束声呐
multibeam sounding sweeping 多波束测探扫海
multibeam sounding system 多波束测探系统
multi-camera system 多相机组合
multichannel 多通道
multichannel filtering 多通道滤波
multichannel receiver 多通道接收机
multi-class 多类
multidimensional random variable 多维随机变量
multi-directional 多角度的
multi-fractal 多重分形
multilane highway 多车道高速公路
multilayer organization 多层结构
multimedia map 多媒体地图
multipath 多路径
multipath effect 多路径效应
multipath error 多路误差
multiple scattering, mutli-scattering 多次散射
multiple-span girder 多跨梁

multiple-span structure	多跨结构
multiple-story	多层(建筑)
multiple-structure	多层结构
multiplexer	多倍仪
multiplexing channel	多路复用通道
multiplexing receiver	多路复用接收机
multiplication	乘法
multiplication constant	乘常数
multiplication law (of probability)	概率乘法定律
multiplier	乘数,乘式
multiply	乘
multi-point matching	多点影像匹配
multi-point piezometer	多点压力(压强)计
multi-purpose cadastre	多用途地籍
multiresolution	多分辨率
multiresolution analysis	多分辨率分析
multiresolution model	多分辨率模型
multiresolution segmentation	多尺度分割
multiscale	多尺度
multiscale analysis	多尺度分析
multiscale morphology	多尺度形态学
multiscale representation	多尺度表达
multiscale transform	多尺度变换
multisensor	多传感器
multi-source information fusion	多源信息融合
multi-source remote sensing data	多源遥感数据
multispectral camera	多谱段摄影机
multispectral image	多光谱影像

multispectral imagery	多谱段图像
multispectral photography	多谱段摄影
multispectral remote sensed image	多光谱遥感图像
multispectral remote sensing	多谱段遥感
multispectral scanner (MSS)	多谱段扫描仪
multispectral texture	多光谱纹理
multistage rectification	多级纠正
multistation photogrammetry	多摄站摄影测量
multi-story building	多层建筑
multi-structure elements	多结构元
multitemporal	多时相
multitemporal analysis	多时相分析
multitemporal composition	多时相组合法
multitemporal image management	多时相影像数据管理
multitemporal images	多时相影像
multitemporal remote sensing	多时相遥感
multitemporal SAR image	多时相 SAR 影像
multithread	多线程
multiuse building	多用途建筑物
multi-value	多值的
multi-year mean sea level	多年平均海面
municipal architecture	城市建筑
municipal area	市区
municipal center	市区中心
municipal drainage	城市排水
municipal extension	城市扩建
municipal heating	城市供热
municipal planning	城市规划

municipal road	城市道路
municipal water suply	城市供水
municipality	市政当局，自治市
Munsell color system	蒙塞尔色系
mural map	挂图
mutually disjoint	互不相交

N

nadir	天底,最低点
nadir observation	天底观测
nanometer	纳米(毫微米),10^{-9}米
nanophotogrammetry	电子显微摄影测量
narrow band multispectral	窄带多光谱
narrow beam echo sounder	窄束回声测深仪
narrow correlator	窄相关器
narrow lane	窄巷
national atlas	国家地图集
national basic gravity network	国家重力基本网
national fundamental geographic information system	国家基础地理信息系统
national geomatics center of China (NGCC)	中国国家基础地理信息中心
national GPS control network	国家 GPS 大地控制网
national hazards	自然灾害
national height datum 1985	1985 国家高程基准
national horizontal control network	国家平面控制网
national land planning	国土规划
national leveling network	国家水准网
national reference frame	国家参考框架
national spatial data infrastructure (NSDI)	国家空间数据基础设施

national technical committee for geographical information, CSBTS/TC 230	全国地理信息标准化技术委员会
national vertical control network	国家高程控制网
national vertical datum	国家高程基准
natural draft	自然通风
natural drainage	自然排水
natural environment	自然环境
natural light	自然光
natural logarithm	自然对数
natural number	自然数
natural terrain	天然地形
nature of the coast	海岸性质
nature reserve	自然保护区
nautical almanac	航海天文历
nautical chart	航海图
nautical Mile	海里（1 海里 = 1852 米）
navigation	导航
navigation channel chart	航道图
navigation chart	导航图
navigation message	导航电文
navigation obstruction	航行障碍物
navigation of aerial photography	航摄领航
navigation star	航用星
navigation station	导航台
navigation station location survey	导航台定位测量
NAVSTAR	导航卫星测时测距

N

Navy Fleet Numerical Oceanographic Center	美国海军海洋学数字数据中心
Navy Hydrographic Office	海军海道测量局
navy navigation satellite system (NNSS)	海军导航卫星系统
neap tide	小潮
near infrared	近红外
near vertical photography	近似垂直摄影
necessary condition	必要条件
negative	负,负片
negative angle	负角
negative image	阴像
negative integer	负整数
negative number	负数
negative vector	负矢量
neighbor identification	邻域识别
neighborhood	邻域
neighborhood method	邻元法
nerve cell	神经元
net area	净面积
net line	建筑线,红线
net of radio satellites	射电卫星空间网
net structure	网状结构
network	网,网络
network adjustment	网平差
network analysis	网络分析
network link	网络联结
network of compass traverse	罗盘仪导线网

N

network RTK	网络 RTK
neural network	神经网络
new chart	初版海图
new construction	新建工程
new edition chart	新版海图
newly-built district	新区
Newton's law of universal gravitation	牛顿万有引力定律
NGCC	中国国家基础地理信息中心
NMEA	导航数据格式
NNSS	海军导航卫星系统
noise	噪声
noise elimination	消声
noise reduction	噪声降低
noise suppressing	噪声抑制
noise-measurement hydrophone	噪声测量水听器
nominal accuracy	标称精度
nominal load	额定荷载
noncentrality parameter	非中心参数
non-collinear	不共线
non-contact method	非接触法
non-conventional equipment	非常规仪器
nondecreasing function	非减函数
non-deterministic function	非确定性函数
non-geodetic techniques	非大地测量技术
nonharmonic analysis of tide	潮汐非调和分析
nonharmonic constant of tide	潮汐非调和常数

non-isothermal pixel	非同温像元
nonlinear	非线性
nonlinear cooling technique	非线性退化技术
nonlinear diffusion	非线性扩散
non-linear math model	非线性数学模型
nonlinear perturbation	非线性摄动
non-metric camera	非量测相机
non-negative	非负的
non-negative matrix	非负矩阵
non-parametric model	非参数模型
nonsingular	满秩的,非奇异的
nonsingular matrix	非奇异阵
nonsingular square matrix	非奇异方阵
non-spherical particle	非球形粒子
nonstationary process	非平稳过程
nonstationary series	非平稳序列
nonstationary stochastic model	非平稳随机模型
nontopographic photogrammetry	非地形摄影测量
non-uniform settlement	不规则沉降
non-zero eigenvalue	非零特征值
non-zero element	非零元素
non-zero vector	非零向量
norm	范数,标准,规范
normal	垂直的,正态的,正常的
normal curve	正态分布曲线
normal distribution	正态分布,常态分布
normal equation	法方程

N

normal error distribution curve	正态误差分布曲线
normal gravitational potential	正常引力位
normal gravity	正常重力
normal gravity field	正常重力场
normal gravity formula	正常重力公式
normal gravity plumb line	正常重力线
normal gravity potential	正常重力位
normal height	正常高
normal level ellipsoid	正常水椭球,水准椭球
normal potential	正常位
normal projection	正轴投影
normal random variable	正态随机变量
normal section	法截面,法截线
normal section azimuth	法截线方位角
normal size	正常尺寸,标准尺寸
normal to curve	曲线的法线
normal vector	法向量
normal-dynamic height	正常动力高
normality of observations	观测值的正态性
normalize	正规化
normalized difference vegetation index(NDVI)	归一化植被指数
normalized fluorescence height	归一化荧光高度
normalized form	标准型
normalized polarization index (NDPI)	归一化极化指数
normal-orthometric height	正常正高
north arrow	指北针

N

north celestial pole	北天极
north direction	北方向
north ecliptic pole	北黄极
north geographical pole	地理北极
north geomagetic pole	地磁北极
north magnetic pole	磁北极
north pole	北极
northern latitude	北纬
north-finding instrument, polar finder	寻北仪器
northing	北
north-seeking	寻北
notice to mariners	航海通告
Novatel	加拿大诺瓦泰
nuclear magnetic resonance (NMR)	核磁共振计算机断层扫描仪
null (zero) hypothesis	零假设
null matrix	零矩阵
null vector	零向量
numerator	分子
numerical accuracy	数值精度
numerical integration	数值积分法
numerical simulation	数值模拟
nutation	章动

O

object contrast	景物反差
object identification, object recognition	目标识别
object location	目标定位
object oriented	面向对象
object region	目标区域
object resolution	目标分辨率
object space coordinate system	物空间坐标系
object space image matching	物方影像匹配
object spectrum characteristics	地物波谱特性
object structure	结构特征
object tracking	对象跟踪
objective angle of image field	像场角
objective function	目标函数
oblique	倾斜的
oblique aerial photograph	倾斜航空摄影像片
oblique observation	倾斜观测
oblique photograph	倾斜摄影像片
oblique photography	倾斜摄影
oblique projection	斜轴投影
observation base	观测基线
observation equation	观测方程
observation error	观测误差
observation of slope stability	边坡稳定性观测

observation plan 观测方案
observation set 测回
observation station of surface movement 地表移动观测站
observation, observed value 观测值
observational astronomy 观测天文学
observatory 观象台,天文台
observed gravity 观测重力值
observer's meridian 测站子午圈
obtuse angle 钝角
obtuse angle triangle 钝角三角形
occlusion 影像遮蔽
occlusion compensation 遮蔽补偿
ocean bathymetric topography 大洋地形图
ocean bottom gravimeter 海底重力仪
ocean buoy 海洋浮标
ocean chart 大洋图
ocean colour remote sensing, remote sensing of ocean color 水色遥感
ocean current 洋流
ocean disaster 海洋灾害
ocean hydrological chart 海洋水文图
ocean island 海洋岛
ocean remote sensing 海洋遥感
ocean topography 海面地形
ocean wave 海浪
oceanic basin 海盆
oceanic observation 海洋观测

oceanic plateau	海底高原
oceanic sailing chart	远洋航行图
Oceanographic Office	海军海洋局
oceanographic survey	海洋调查
oceanography	海洋学
octagon	八边形
octahedron	八面体
octree	八杈树
ODBC	开放数据库互联
odd function	奇函数
odd number	奇数
odd-even embedding	奇偶嵌入法
odometer	计程仪,里程表
off diagonal element	非对角线元素
offset	偏移量
offset printing	胶印
offshore oil well	海上油井
offshore platform	海上平台
offshore sailing chart	近海航行图
offshore survey	近海测量
OGC	开放式地理信息系统协会
oil extraction	采油
old architecture survey	古建筑测绘
omnidirectional	全向的,无定向的
omnidirectional antenna	全向天线
one-sided test	单边检验
online access	在线访问

O

on-line aerophotogrammetric triangulation	联机空中三角测量
on-site installation	现场安装
On-The-Fly (OTF)	在航
open database connectivity (ODBC)	开放数据库互联
Open Geodata Interoperability Specification	开放式地理数据互操作规范
open geographic information system, Open GIS	开放式地理信息系统
Open GIS Consortium(OGC)	开放式地理信息系统协会
open interval	开区间
open traverse	支导线
opencast mining plan	露天采矿图
opencast survey	露天矿测量
operating cost	运行成本,使用费
operation	操作,运算
operation and maintenance	运营和维护
operational life	使用年限
opposite angle	对角
opposite side	对边
optical alignment	光学准直
optical axis	主光轴
optical compensator	光学补偿器
optical condition	光学条件
optical correlation	光学相关
optical density	光学密度
optical fiber	光纤

O

optical flow	光流
optical image	光学影像
optical level	光学水准仪
optical plummet	光学对中器
optical point seismometer	光点地震仪
optical projection	光学投影
optical rangefinder	光学测距仪
optical scanner	光学扫描仪
optical sensor	光学传感器
optical theodolite	光学经纬仪
optical train	光学系统
optical transfer function(OTF)	光学传递函数
optimal design of geodetic networks	大地测量控制网优化设计
optimal hyperplane	最优超平面
optimal network design	控制网优化设计
optimal solution	最优解
optimal threshold	最优阈值化
optimum configuration	参数配置
optimum design	最优设计
optimum wave-band combination	最优波段组合
orbit	轨道
orbit determination	定轨
orbit parameters	轨道参数
orbital precession	轨道岁差
Orbview 4	轨道观测卫星系统，美国轨道科学公司(Orbimage) 2001 年发射

O

order of survey	测量等级
ordinary differential equation	常微分方程
ordinate	纵坐标
ordinate axis	纵坐标轴,Y轴
orientation	定向
orientation of reference ellipsoid	参考椭球定位
orientation parameters	定向参数
orientation survey	定向测量
orienteering map	定向运动地图
origin	原点
origin of control network	控制网原点
origin of longitude	经度起算点
ortho rectification	正射纠正
orthocentre	垂心
orthochromatic film	正色片
orthogonal	正交
orthogonal transformation	正交变换
orthogonal wavelet transform	正交小波变换
orthogonality	正交性
orthographic perspective	正射透视,正射透视图
orthographic projection	正射投影
orthography of geographical name	地名正名
orthoimage	正射影像
orthometric height	正高
orthophoto	正射像片
orthophoto map	正射影像地图
orthophoto stereomate	立体正射像片

orthophoto technique	正射影像技术
orthophotograph	正射摄影
orthophotography	正射投影
orthophotomosaic	正射像片镶嵌图
orthostereoscopy	正立体效应
oscillation	振动
oscillatory convergence	振动收敛性
outage	不可用
outboard antenna	舱外天线
outer core	外核
outlet	出水口,电源插座
outlet pipe	排水管
outlier	异常值,粗差
outline map [for filling]	填充地图
output box	输出框
output signal	输出信号
outstanding point	明显地物点
over crossing	人行天桥
overall (global) precision criterion	总体精度准则
overall (global) precision estimates	总体精度估值
overall dimension	总尺寸
overall height	总高度
overall length	总长度
overall perspective	全景图
overall planning	总体规划
overall urban layout	城市整体布局
overhanging beam	悬臂梁

overlap	重叠度,交叠
overlapping area correction	接边纠正
overlapping image method	重叠影像法
overlay	叠加
overlay analysis	叠置分析
own weight	自重
ownership	所有权,拥有权

P

pacing	步测,定步
panchromatic	全色的
panchromatic film	全色片
panchromatic image	全色影像
panchromatic infrared film	全色红外片
panel	仪表盘,面板
panoramic	全景的
panoramic camera	全景摄影机
panoramic distortion	全景畸变
panoramic photography	全景摄影
pantograph	缩放仪,缩图器,缩放
parabola	抛物线
parabolic curves	抛物曲线
paraboloid	抛物面
parallax	视差
parallax in altitude	高度视差
parallax in cross-hairs, parallax in reticule	十字丝视差
parallel algorithm	并行算法
parallel channel receiver	并行通道接收机
parallel circle	纬圈,平行圈
parallel force	平行力
parallel lines	平行(直线)
parallel masks	并行模板

P

parallel processing	并行处理
parallel ray approximation	平行射线近似
parallel-averted photography	等偏摄影
parallelepiped	平行六面体
parallelogram	平行四边形
parameter	参数,参变量
parameter constrained adjustment	附条件参数平差,附条件间接平差
parametric adjustment	参数平差,间接平差
parametric equation	参数方程
parametric model	参数模型
parcel	宗地
parcel map	宗地图
parcel survey	地块测量
partial autocorrelation function	偏自相关函数
partial least squares regression	偏最小二乘回归
partial sum	部分和
partial tide	分潮
particle accelerator survey	粒子加速器测量
particular map	特种地图
particular solution	特解
partition	分割,划分
partitioned matrix	分块矩阵
pass point	加密点
passive microwave	被动微波
passive microwave remote sensing	被动微波遥感
passive microwave sensor	被动微波遥感传感器
passive positioning system	被动式定位系统

passive remote sensing	被动式遥感
passive sensor	被动式传感器
passive sonar	被动声呐
patch	斑点,小块地
path	路径
path finding	路径查找
pattern analysis	模式分析
pattern element	图形要素
pattern language	模式语言
pattern plate	模板
pattern recognition	模式识别
pattern recognition of remote sensing	遥感模式识别
pattern texture projector	图形纹理投影器
P-Code	P 码,精码
PDF	概率密度函数
PDOP	位置精度因子
peak value method	峰值方法
pedestian bridge	人行天桥
pedestian crossing	人行横道
pedestian street	步行街
peel-coat film	撕膜片
peg,stake	桩
pelagic sailing chart	远海航行图
pelagic survey	远海测量
pendulum	摆
pendulum instrument	摆仪
pentaprism	五角棱镜

perceived model	视模型
percent	百分之(%)
percentage	百分率
perception law	感知规律
perceptual grouping	类别视觉感受
perigee	[天]近地点
perihelion	[天]近日点
perimeter	周长,周界
period	周期
period of design	设计年限
periodic error	周期误差
periodic function	周期函数
periodic perturbation	周期摄动
periodogram	周期图
permanent building	永久建筑
permanent geodetic marker	永久性测量标
permanent population	常住人口
permanent structure	永久性结构(建筑物)
permanent station	永久测站,埋石测站,埋石点
permissible load	容许荷载
permission notes foe land use	建设用地规划许可证
permutation	排列
perpendicular	垂线,垂直(于)
perpendicular bisector	垂直平分线
personal error	人为误差
perspective	透视,透视图,透视的
perspective centre	透视中心

perspective projection	透视投影
perspective traces	透视截面法
perspective view	透视图
perturbation	摄动
perturbed motion of satellite	卫星受摄运动
petroleum exploration survey	石油勘探测量
phase	相位,相
phase adjustment	分组平差
phase ambiguity	相位模糊度
phase ambiguity resolution	相位模糊度解算
phase compensator	相位补偿器
phase cycle	相位周
phase cycle value	相位周值
phase drift	相位漂移
phase lock loop(PLL)	锁相环
phase shift	相移
phase stability	相位稳定性
phase transfer function(PTF)	相位传递函数
phase unwrapping	相位解缠
phase velocity	相速
phase-smoothed pseudo-range	相位平滑伪距
photo coordinate system	像平面坐标系
photo interpretation	像片判读
photo mosaic	像片镶嵌
photo nadir point	像底点
photo orientation elements	像片方位元素
photo rectification	像片纠正
photo scale	像片比例尺

photo theodolite, camera transit	摄影经纬仪
photo tilt	像片倾斜
photo, photograph	像片
photocell	光电管,光电池,光电元件
photoelectric sensor	光电遥感器
photogrammetric coordinate system	摄影测量坐标系
photogrammetric distortion	摄影测量畸变差
photogrammetric instrument	摄影测量仪器
photogrammetric interpolation	摄影测量内插
photogrammetric robot	摄影测量机器人
photogrammetrist	摄影测量员
photogrammetron	摄影测量机
photogrammetry	摄影测量学
photogrammetry baseline	摄影测量基线
photograph	摄影,像片
photographic	摄影的
photographic astrometry	摄影天体测量学
photographic baseline	摄影基线
photographic detail	影像细部,图像细部
photographic principal distance	摄影主距
photographic processing	摄影处理
photographic quality	摄影质量
photographic scale	摄影比例尺
photography	摄影学
photomap	影像地图
photomicrography	显微摄影
photoplan	像片平面图

physical enviroment	自然环境
physical geodesy	物理大地测量学
physical map	自然地图
physical oceanography	物理海洋学
physical surface	自然表面
physics	物理学
picometer	皮米(微微米),10^{-12}米
pictogram	象形图
picto-line map	浮雕影像地图
picture element/ pixel	像素
picture format	像幅
pie chart	饼图
pier	桥墩
pier centre	桥墩中心
pier construction	桥墩施工
piezoelectric transducer	压电换能器
piezometer	压力(压强)计
pile	桩
pile foundation	桩基础
pillar	支柱,压杆
pilot anchorage	引水锚地
pilot atlas	引航图集
pilot chart	引航图
Pinnacle	JAVAD GPS 接收机随机软件
pipe alignment	管道定线
pipe survey	管道测量

pipeline	管道,管线
pipeline network	管网
pipeline survey	管线测量
piping system	管道系统
pitch	纵摇
pitometer log	水压计程仪
pivot	支点
pixel	像元,像素
pixel location	像素定位
pixel unmixing	像元分解
place name	地名
place-name database	地名数据库
plan study	方案研究
plan view	平面图,鸟瞰图
planar grid	平面格网
plane	平面
plane curve location	平面曲线测设
plane figure	平面图形
plane surveying	平面测量,平面测量学
plane triangle	平面三角形
planetary geodesy	行星大地测量学
planetology	行星学
planimeter	求积仪
planimeter method	求积仪法
planimetric map	平面图,鸟瞰图
planning criteria	规划标准
planning map	规划图

planning permission 规划许可
planning procedure 规划程序
planning survey 规划调查
plantary astronomy 行星天文学
plantary precession 行星岁差
plate correction 层间改正
plate tectonics 板块构造学
platen 平台
plot 绘图,展绘
plot ration 容积率
plotter 绘图机
plotting file 绘图文件
plotting table 绘图桌,量测台
plumb aligner 垂准仪,铅垂仪
plumb bob 垂球
plumb line 铅垂线
point cloud 点云
point coordinate positioning 极坐标定位,距离方位定位
point estimate 点估计
point light source 点光源
point mode 点方式
point of Aries, vernal equinox 春分点
point of intersection 交点
point of secant 割点
point of suspension 悬点
point of tangency 切点
point positioning 单点定位

P

point sample	点样本校验
point spread function(PSF)	点扩散函数
point symbol	点状符号
point target	点目标
pointing accuracy	照准精度,目标指示精度
Poisson distribution	泊松分布
polar axis,pole axis	极轴
polar comparator	极坐标量测仪
polar coordinate	极坐标
polar coordinate system	极坐标系统
polar motion	极移
polar pantograph	极坐标缩放仪
polar region	极地
polar stereographic projection	极球面投影
polar triangle	极三角形
polarimetric analysis	极化分析
polarimetric radar	极化雷达
polarimetric SAR interferometry	极化干涉
polarized light	偏振光
polarized reflection	偏振反射
pole	极
polyconic projection	多圆锥投影
polyfocal projection	多焦点投影
polygon	多边形
polygon overlay	多边形叠置
polygonal map	多边形地图
polyhedron	多面体

polynomial	多项式
polynomial adjustment	多项式平差
polynomial equation	多项式方程
polynomial regression	多项式回归
poor engineering	劣质工程
population	总体
population density	人口密度
population distribution	人口分布
population growth rate	人口增长率
population map	人口地图
population mean	总体平均(值)
population structure	人口结构
Porro-Koppe principle	波罗-科普原理
port	港口
portable echo sounder	便携式回声测深仪
portable sonar	便携声呐
portable tide gauge	便携式验潮仪
portal	入口
POS	定位与定向系统,POS 系统
position	位置
position and attitude	位置和姿态
position and orientation system (POS)	定位与定向系统,POS 系统
position dilution of precision (PDOP)	位置精度因子
position format	位置格式
position function	位置函数,坐标函数

P

position sensitive detector(PSD) 位置传感探测器
position vector 位置向量
positional accuracy 位置精度
positional astronomy 方位天文学
positioning interval 定位点间距
positioning mark 定位标志
positioning, position fix 定位
positive 正片,正
positive image 正像,正片
positive index 正指数
positive integer 正整数
positive number 正数
positive positioning system 主动式定位系统
posterior probability 后验概率
posteriori 后验
posteriori variance factor 后验方差因子
post-glacial rebound 冰后回弹
post-processed differential correction 后处理差分改正
post-processed GPS 后处理 GPS
post-processed kinematic(PPK) 后处理动态
postulate 假设,假定
potential energy 势能
potential field 势能场
power 功率
power generation 发电
power outlet 插座,插孔
power spectrum 功率谱

power transmission line survey 输电线路测量
PPK 后处理动态
PPP 精密单点定位
PPS 精密定位服务
practical astronomy 实用天文学
practical salinity 实用盐度
practical salinity scale 实用盐标
pre-amplifier 前置放大器
precession 岁差,进动
precession angular velocity 进动角速度
precession of equinox 分点岁差
precessional motion 进动
precise alignment 精密准直
precise alignment observation 精密准直测量
precise Code(P Code) 精码
precise engineering control network 精密工程控制网
precise engineering survey 精密工程测量
precise ephemeris 精密星历
precise level 精密水准仪
precise leveling 精密水准测量
precise plumbing 精密垂准
precise point positioning (PPP) 精密单点定位
precise positioning service (PPS) 精密定位服务
precise ranging 精密测距
precise traversing 精密导线测量
precise triangulation 精密三角测量
precision 精度
precision agriculation (farming) 精细农业,精细农作

precision analysis	精度分析
precision estimation	精度估计
prediction of deformation	变形预测
preliminary drawings	初步设计图
preliminary estimate	设计概算
preliminary orientation	粗略定向
preliminary survey	初测,勘测
premarking	预设标志
pre-press proof	预打样图
preprinted symbol	预制符号
pre-processing	预处理
prescribed course	规定航向
pre-selection sample	样本预选取
presensitized plate	预制感光板,PS 版
present landuse map	土地利用现状图
pressure altimeter	气压高程计,气压测高仪,高差仪
pressure autorecording tide gauge	压力式自计验潮仪
pressure gauge	压力验潮仪
prestressed concrete	预应力混凝土
preview	预览
pre-warning system	预警系统
priliminary triangle	概算三角形
primary	第一位,首级
primary control network	首级控制网
primary graphic elements	基本图形元素
primary gravity network	首级重力网
primary scale	基本比例尺,主比例尺,游标比例尺

primary triangulation net	一等三角网
prime	素
prime meridian	本初子午线
prime number	素数
prime vertical circle	卯酉圈,东西圈
prime vertical section	卯酉面
principal angle	主角
principal axis	主轴
principal component analysis (PCA)	主成分分析,主分量分析
principal component pixels decomposition	主成分像元分解
principal component transformation	主分量变换
principal distance	主距
principal distance of camera	摄影机主距
principal distance of photo	像片主距
principal epipolar line	主核线
principal epipolar plane	主核面
principal line [of photograph]	像主纵线
principal network	主网
principal plane [of photograph]	主垂面
principal point of autocollimation (PPA)	自准直主点
principal point of photograph	像主点
principal traverse	基本导线,主导线
principal vanishing point	主合点
principle of geometric reverse	几何反转原理
printer lens	照相制版镜头

P

printing plate	印刷版
prior knowledge	先验知识
priori	先验
priori probability	先验概率
priori variance factor	先验方差因子
prism	棱镜
probability	概率
probability decision function	概率判别函数
probability density function (PDF)	概率密度函数
probability distance	概率距离
probability distribution	概率分布
probability generating function	概率母函数
probability level	概率水平
probability relaxation algorithm	概率松弛法
probability space	概率空间
probable error	概然误差,或然误差
procedure for approval of urban plan	规划审批程序
process box	处理框
product	乘积,积
product set	积集
profile	纵断面
profile diagram	纵断面图
profile extraction	断面提取
profile leveling	纵断面水准测量
profile survey	纵断面测量
prognostic map	预报地图
program-controlled instrument	程控仪器

P

progressive image retrieval	渐进式图像检索
prohibited area	禁航区
project design	项目设计
project engineer	项目工程师
project management	项目管理
project planning	项目规划
project profile	工程概况
project site	建筑现场,工地
project supervision	项目监督
projected scale	投影尺度,投影比例尺
projectile motion	抛射运动
projection	投影(映)
projection distortion	投影变形
projection equation	投影方程
projection lines	投射线
projection plane	投射平面
projection transformation	投影变换
projector	投影器
proof by contrapositive	反证法
proper scale	成图比例尺,规定比例尺
property	房地产,权属
property survey	不动产测量,权属测量
property line	建筑红线
property line survey	建筑红线测量
proportion	比例

proportional	成比例
proportional error	比例误差
proportional scale	比例尺
prospecting baseline	勘探基线
prospecting line survey	勘探线测量
prospecting network layout	勘探网测设
provisional chart	临时海图
pseudo-color	伪彩色
pseudo-color image	伪彩色图像
pseudolite	伪卫星
pseudo-random noise (PRN)	伪随机噪声
pseudo-random noise code	伪随机噪声码
pseudo-range	伪距
pseudo-range measurement	伪距测量
pseudostereoscopy	反立体效应
psychological rescue	心理救助
psychrometer	干湿球温度表
public building	公共建筑
public works	市政工程
public works engineering survey	市政工程测量
pulley	滑轮
push-broom sensor	推扫式传感器
pylon	桥塔
pyramid principle	角锥体法

Q

quadrant	象限
quadrilateral	四边形，四边地
qualitative perception	质量感
quality control and validation	质量控制与评估
quality evaluation	质量评价
quality of aerophotography	航摄质量
quantitative evaluation	量化评价
quantitative perception	数量感
quantitative remote sensing	定量遥感
quantitative remote sensing model	定量遥感模型
quantization, quantizing	量化
quasi-2D control field	二维控制场
quasi-dynamic height	似动力高
quasi-geoid	似大地水准面
quasi-geoid height	似大地水准面高
quasi-stable adjustment	拟稳平差

R

radar 雷达

radar altimeter 雷达测高仪

radar imagery 雷达图像

radar interferometry 雷达干涉测量

radar overlay 雷达覆盖区

radar ramark 雷达指向标

radar responder 雷达应答器

radar scene matching 雷达影像匹配

radarclinometry 雷达测角

radargrammetry 雷达图像测量学,雷达图像测量技术

RADARSAT 加拿大发射的雷达卫星

RadarSat 2 微波遥感卫星,1995年发射

radial basis function neural networks (RBFN) 径向基函数神经网络

radial basis function(RBF) 径向基函数

radial component 沿径分量

radial distortion 径向畸变

radial parallax 径向视差

radian 弧度

radiance 辐射亮度

radiation compensation 辐射补偿

radiation correction	辐射校正
radiation ratio of signal to noise	辐射信噪比
radiation scaling	辐射度量
radiation sensor	辐射遥感器
radiation transfer model	辐射传输模型
radiational tide	辐射潮
radiative calibration	辐射定标
radiative transfer	辐射传输
radiative transfer equation	辐射传输方程
radio altimeter	无线电测高仪
radio beacon	无线电指向标
radio frequency (RF)	射频
radio navigational warning	无线电航行警告
radio positioning	无线电定位
radio satellite	射电卫星
radio triangulation	无线电大地测量,雷达三角测量
radio wave	无线电波
radiometer	辐射计,射线探测仪
radiometer scatterometer	散射计辐射计组合
radiometric alteration	色彩平衡
radiometric correction	辐射改正
radiosonde	探空
radius of curvature in prime vertical	卯酉圈曲率半径
radius of curvature of normal section	法截线曲率半径
radius of gyration	回转半径

R

radius of parallel 平行圈半径
radius of railway curve 铁路曲线半径
radius of vertical curve 竖曲线半径
radius, radii 半径
railroad engineering survey 铁路工程测量
random 随机
random error 偶然误差
random experiment 随机试验
random number 随机数
random sample 随机样本
random sampling 随机抽样
random variable 随机变量
range estimation 距离估测
range image 距离图像
range positioning system 测距定位系统
range resolution 距离分辨率
rangefinder 测距仪
range-only radar 测距雷达
range-range positioning 圆-圆定位,距离-距离定位
rank 秩
rank defect adjustment 秩亏平差
rank of a matrix 矩阵的秩
raster data 栅格数据
raster plotting 栅格绘图
raster structure 栅格结构
rasterization 栅格化
raster-to-vector conversion 栅格-矢量转换

rate of change	变率
rate of convergence	收敛率
rate-distortion	比特-失真率
ratio	比
ratio analysis	比值分析
ratio enhancement	比值增强
ratio scaling	比例量表
ratio transformation	比值变换
rational function	有理函数
rational number	有理数
rationalization	有理化
raw data	原始数据
ray diagram	声线图
rayleigh reflectance	瑞利反射率
rayleigh scattering	瑞利散射
reaction (force)	反作用(力)
reading accuracy	读数精度
reading accuracy of sounder	测深仪读数精度
reading scale	刻画板
real aperture radar	实孔径雷达
real aperture radar image	实孔径雷达图像
real axis	实轴
real coding genetic optimize algorithm	实码遗传优化算法
real estate	房地产
real image matching	实时影像匹配
real number	实数
real part	实部

R

real root 实根
real time kinematic (RTK) 实时动态定位
real time rectification 准实时纠正
real vector space 真矢量空间
real-aperture radar 真实孔径雷达
real-time 实时
real-time compress 实时压缩
real-time DGPS 实时差分 GPS
real-time differential correction 实时差分改正
real-time photogrammetry(RTP) 实时摄影测量
real-time processing 实时处理
real-time visualization 实时可视化
realty 房地产
realty industry 房地产业
reapparition of color signal 色彩信号复现
receiver 接收机
receiver antenna 接收机天线
receiver array 接收机阵列
receiver autonomous integrity monitoring(RAIM) 接收机自主完好性监测
receiving center 接收中心
rechargeable battery 充电电池
reciprocal 倒数
reciprocal observation 对向观测
reciprocal transducer 互易换能器
reciprocal trig leveling 对向三角高程测量
reclaimation survey 复垦测量
recognition 识别

recognition effect	识别效果
recollimation	重新照准
recommended route	推荐航线
reconnaissance	勘测
reconnaissance diagram	勘测图
reconnaissance report	勘测报告
reconnaissance survey	初测,勘测
record format	记录格式
recorder of soundings	测深仪记录器
recording paper of sounder	测深仪记录纸
rectangle	长方形
rectangular block	长方体
rectangular coordinate	直角坐标
rectangular coordinate system	直角坐标系统
rectangular distribution	矩形分布
rectangular grid	直角坐标网
rectangular plane coordinate	平面直角坐标
rectangular subdivision	矩形分幅
rectification	纠正,整流
rectifier, transformer	纠正仪
recurrence formula	递推公式
recurring decimal	循环小数
red edge parameter	红边参数
reduced distance	改化距离
reduced gravity	改化后的重力值
reduced latitude	归化纬度
reduced longitude	改化经度
reducing color printing	减色印刷

reduction	归算
reduction dimension	降维
reduction for polar motion	极移改正
reduction formula	归约公式
reduction of sounding	测深归算
reduction to geodesic	截面差改正
redundancy	冗余度
redundancy number	多余观测数
redundant code	冗余码
redundant information	冗余信息
redundant observation	多余观测
reef	暗礁
re-enforced concrete	预应力混凝土
reference axis	坐标轴,基准轴
reference azimuth	球面方位角
reference base stations	参考基站
reference central meridian	基准中央子午线,基准中央经线
reference data	参考数据
reference datum	参考基准
reference ellipsoid	参考椭球
reference image	参考基准影像
reference line	基准线
reference mark	参考标志
reference plane	参考平面
reference point	参考点
reference receiver	参考站接收机
reference space	参考空间

R

reference spheroid 参考球体,参考椭球体
reference station 参考站
refill 回填,再填
refined Bouguer gravity anomaly 精化布格重力异常
reflectance 反射,反射比
reflectance model 反射率模型
reflectance retrieval 反射率转换
reflectance spectrum 反射光谱
reflecting stereoscope 反光立体镜
reflectivity 反射性,反射率,反射系数
reflector 反射器
reflectorless total station 无反射全站仪
reflexive relation 自反关系
refraction correction 折光差改正
region 区域
region filling 区域填充
region growth 区域增长算法
region merging 区域合并
region of acceptance 接收区域
region of convergence 收敛区域
region of rejection 拒绝区域
regional analysis 区域分析
regional atlas 区域地图集
regional chart 海区图
regional geological map 区域地质图
regional geological survey 区域地质调查

Regional Hydrographic Commission	区域海道测量委员会
regional planning	区域规划
regionalization map	区划地图
region-growing	区域增长
registered architect	注册建筑师
regular grid DEMs	规则格网数字高程模型
regular square grid	规则格网模型
regularization	规则化
regularized geoid	调整[的]大地水准面
reinforced concret(RC)	钢筋混凝土
reinforced concrete	预应力混凝土
relational matching	关系匹配
relative (point) error ellipse	相对(点位)误差椭圆
relative accuracy	相对精度
relative altitude	相对高度
relative control	相对控制
relative displacement	相对位移
relative error	相对误差
relative frequency	相对频数
relative gravimeter	相对重力仪
relative gravity	相对重力
relative gravity measurement	相对重力测量
relative linear error of traverse	导线相对闭合差
relative maximum	相对极大
relative minimum	相对极小
relative motion	相对运动
relative orientation	相对定向

relative parallax	相对视差
relative position	相对位置
relative positioning	相对定位
relative radiometric correction	相对辐射校正
relative tilt	相对倾角
relative velocity	相对速度
relativistic correction	相对论改正
relativistic effect	相对论效应
relativity	相关性,相对论
relaxation algorithm	松弛算法
reliability	可靠性
reliability design	可靠性设计
relief displacement	投影差
relief feature	地形特征,地形要素
relief map	立体地图
remote controller	遥控器
remote sensing data acquisition	遥感数据获取
remote sensing exploration	遥感勘查
remote sensing for natural resources and environment	资源与环境遥感
remote sensing images from different orbits	异轨遥感影像
remote sensing information source	遥感信源
remote sensing interpretation	遥感判读
remote sensing mapping	遥感制图
remote sensing monitoring	遥感监测
remote sensing platform	遥感平台
remote sensing retrieval	遥感反演

remote sensing sounding	遥感测深
remote sensing(RS)	遥感
remote sensor	遥测传感器,遥感器
renewable energe resource	再生能源
renewal of the cadastre	地籍更新
reorganization	重组织
repeated observation	重复观测
repeated trials	重复试验
repetition method	复测法
replicative symbol	象形符号
representative fraction(RF)	分数比例尺
resampling	重采样
resection	后方交会
reservoir survey	库容测量,水库测量
residential area	居民地
residential density	居住密度
residential district	居民区
residual gravity	剩余重力值
residual sum of squares	残差平方和
residual,residue	残差
resolution	分辨率,分辨力
resolution acuity	视觉分辨敏锐度
resolution enhancement	分辨率增强
resolving power of lens	物镜分辨率
resource demand	资源需求量
resource support system	资源支持体系
restitute	重建,恢复,偿还
restoration	恢复,修复

restricted area	限航区
resultant velocity	合速度
reticule	十字丝
retouching	修版
retrieval method	反演方法
retro-reflective targets(RRT)	回光反射标志
retroreflectors	逆向反射器,反光镜
return beam vidicon camera(RBV)	反束光导管摄像机
return wave	回波
reversal film	反转片
reversal or turning point	逆转点
reversal points method	逆转点法
reverse curve	回头曲线
reverse radiative temperature	逆向辐射温度
reversing current	往复流
revision of topographic map	地形图更新
revolution per minute(PRM)	主机航速
revolution,rotation	旋转
rhombus	菱形
rhumb line sailing	等角航法
ridge estimation	岭估计
ridge line	山脊线
ridge parameter	岭参数
ridge-stein combined estimator	岭-压缩估计组合
Riemannian space	黎曼空间
right angle	直角
right circular cone	直立圆锥
right circular cylinder	直立圆柱

right femur	右边
right of land usage	土地使用权
right of occupancy	居住权
right spherical triangle	球面直角三角形
right-handed coordinate systems	右手坐标系
right-hand-side(R. H. S)	右手边
right-of-way	可通行
rigid body	刚体
rigid frame	刚结构
rigorous adjustment	严密平差
RINEX	GPS 标准数据交换格式
rip current, rip surf	离岸流
rise	海隆
risk management	风险管理
risk warning system	预警系统
river improment survey	河道整治测量
riverbed	河床
river-crossing leveling	跨河水准测量
RMS	均方根
RMSE	均方根误差,中误差
road border	路肩,路边
road clearance	道路净空
road detection	道路探测/检测
road extraction	道路提取
road network	道路网
road searching	道路搜索
robotic (motorized) total station	智能型全站仪,测量机器人

R

robotic system	机器人系统
robust estimation	抗差估计,稳健估计
robust statistics	稳健统计学
robustness	稳健性
rock bolting	锚固
rock classification	岩性分类
rod	标尺
ROI	样区
roll	旁向倾角,横摇
roller filtering compensator	涌浪滤波补偿器
roller wheel algorithm	滚轮算法
rolling terrain	丘陵地
roof station	顶板测点
root mean square (RMS)	均方根
root mean square error(RMSE)	均方根误差,中误差
rotary current	回转流
rotating aerial	回转天线,旋转天线
rotating axis of the gyro	陀螺旋转轴
rotating mirror	旋镜,旋转镜,回转反射镜
rotation axis	旋转轴,公转轴
rotation invariant	旋转不变
rotation matrix	旋转矩阵
rotation parameters	旋转参数
rotational angular velocity of the earth	地球自转角速度
rough logic	粗糙逻辑
rough set	粗集

rough set theory	粗糙集理论
route	航线,线路
route alignment	公路定线
route analysis	路径分析
route engineering survey	线路工程测量
route leveling	线路水准测量
route plan	线路平面图
route selection	选线
route stationing	线路测设
route survey	线路测量
rover	流动站
roving receiver	流动接收机
row vector	行向量
RS	遥感
RS-232	数据通信串口协议
RTK	实时动态定位
RTP	瑞士的实时摄影测量系统
ruling	网线
run length encoding	行程编码,游程编码
rural planning	乡村规划

S

SA	选择可用性
safe clearance	安全净空
safety of network	网络安全
safty distance	(建筑物)安全间距,安全距离
safty exit	安全出口
safty strip	安全区,安全带
sag	下垂,下陷,垂度
sailing chart	航行图
sailing directions	航路指南
salinity of sea water	海水盐度
salinometer	盐度计
sample	采样
sample space	样本空间
sampling	采样,抽样
sampling density	采样密度
sampling distribution	抽样分布
sampling for training areas	训练区采样
sampling interval	采样间隔
sampling rate	采样率
sampling theorem	抽样定理
sand map	沙盘地图
SAR	合成孔径雷达
SAR complex image	SAR 复数图像

S

satellite town	卫星城
satellite acceleration sensor	星载加速度传感器
satellite altimeter	卫星高度计
satellite altimetry	卫星测高
satellite altitude	卫星高度
satellite attitude	卫星姿态
satellite clock	卫星钟
satellite configuration	卫星构形
satellite constellation	卫星星座
satellite Doppler positioning	卫星多普勒定位
satellite Doppler shift measurement	卫星多普勒[频移]测量
satellite geodesy	卫星大地测量,卫星大地测量学
satellite gradiometry	卫星重力梯度测量
satellite image	卫星影像
satellite imagery of push broom	推扫式遥感影像
satellite images	卫星成像
satellite laser ranger	卫星激光测距仪
satellite laser ranging(SLR)	卫星激光测距
satellite oceanography	卫星海洋学
satellite orbit	卫星轨道
satellite orientation	卫星定向
satellite photogrammetry	卫星摄影测量
satellite photography	卫星摄影
satellite positioning	卫星定位
satellite remote sensing fathoming	卫星遥感测深
satellite sensor	星载遥感器

satellite tracking station	卫星跟踪站
satellite-based augmentation system (SBAS)	星基增强系统
satellite-to-satellite tracking(SST)	卫星跟踪卫星技术
saturation	饱和度
SBAS	星基增强系统
SBSM	国家测绘局
scale	比例尺,尺度
scale accuracy	比例尺精度,尺度精度
scale based concurrent matrix	尺度共生矩阵
scale distortion	比例尺变形
scale factor	比例因数
scale of depth	深度比例尺
scale parameter	尺度参数
scale space	尺度空间
scale variation	比例尺变化
scale-space representation	尺度空间表达
scaling	定标,定比例,量测
scaling constant	比例尺常数
scaling down	分解(遥感尺度转换)
scaling effect	尺度效应
scaling of model	模型缩放
scaling up	聚合(遥感尺度转换)
scan angle	扫描角
scan lines	扫描行
scan mode of airborne-based image laser radar	机载激光成像雷达扫描成像模型

scan-digitizing	扫描数字化
scanner	扫描仪
scanning	扫描
scanning electron microscope	扫描电子显微镜
scanning frequency	扫描频率
scanning laser rangefinder	激光扫描测距
scanning mirror	扫描镜
scanning order	扫描次序
scanning trace	扫描轨迹
SCANSAR	扫描雷达成像技术
SCAR	南极科学考察委员会
scattering characteristic	散射特性
scattero radiometer	散射-辐射计
scatterometer	散射计
scatterometry	散射测量
scene matching	景象匹配
scheimpflug condition	交线条件
schematic design	方案设计
schematic design phase	方案设计阶段
scheme	方案,计划
Schuler mean	舒勒平均值
science museum	科技馆
science of ocean wave	海浪学
Scientific Committee on Antarctic Research(SCAR)	南极科学考察委员会
Scientific Committee on Oceanic Research(SCOR)	海洋研究科学委员会
scientific notation	科学计数法

S

scientific terminology	科技术语
SCOP	德国 Stuttgart 大学研制的 DTM 软件
SCOR	海洋研究科学委员会
screen	屏幕
screen map	屏幕地图
screw	螺旋
scriber	刻图仪
scribing	刻绘
SDI	空间数据基础设施
SDTS	空间数据转换标准
sea area bounding line	海区界线
sea coast	海岸
sea current	海流
sea echo	海面回波
sea noise	海洋噪声
sea surface height	海面高
sea surface topography model	海面地形模型
sea surface topography(SST)	海面地形
sea survey	海面观测
sea swell chart	海浪图
seabed feature survey	海底面状探测
seafloor acoustic beacon	海底声标
seafloor gravimeter	海底重力仪
seafloor imaging system	海底成像系统
seafloor slope correction	海底倾斜改正
seafloor topography	海底地形
sea-level datum	海平面基准面

S

sea-level elevation	海拔高程
sealing cavity method	比辐射率封闭测定法
seam line removal	拼接缝消除
seamless contiguity	无缝拼接
seamless database	无缝数据库
seamless registration	无缝镶嵌
search strategy	搜索策略
searching area	搜索区
seasat	海洋卫星
seashore	海滨
seasonal correction of mean sea level	平均海面归算
seasonal correction of sea level	平均海面季节性改正
second derivative	二阶导数
second eccentricity	第二偏心率
second order ordinary differential equation	二阶常微分方程
secondary	第二位,次级
second-order design (SOD): the weight problem	二类设计:观测值权的分配问题
section survey	断面测量
sectional drawing	剖面图,断面图
sectional plane	剖面
sector	扇式
sectorial spherical harmonics	扇球谐函数,扇球调和函数
seismic detector	地震检波器
seismic focus	震源

S

seismic geophysical method	地球物理地震法
seismic intensity	地震强度
seismic magnitude	地震震级
seismic method	地震勘测法
seismic signal processor	地震回放仪
seismic site producer	震源发生器
seismic wave	地震波
seismicity map	地震图
selective availability (SA)	选择可用性
selective endmember	端元变化
selenodesy	月面测量,月面测量学
self organizing network	自组织网络
self-adaptive gradient	自适应梯度
selfcalibrating block adjustment	自检校区域网平差
self-calibration	自检校
self-organization	自组织
self-organizing feature map	自组织特征映射
self-similarity	自相似
semantic extraction	语义提取
semantic information	语义信息
semantics sharing	语义共享
semiarid landscape	半干旱草场
semiautomatic extraction	半自动提取
semicircle	半圆
semiconjugate axis	半共轭轴
semicontrolled mosaic	半控制镶嵌图
semidiurnal tidal current	半日潮流

S

semidiurnal tide	半日潮
semidiurnal tide harbor	半日潮港
semimajor axis	长半轴
semi-major axis of ellipsoid	椭球长半轴,地球长半轴
semiminor axis	短半轴
semi-minor axis of ellipsoid	椭球短半轴,地球短半轴
semi-parametric regression	半参数回归
sensitising, sensitization	感光
sensitivity	灵敏度,感光度
sensitivity design	灵敏度设计
sensitometry	感光测定
sensor attitude	传感器姿态
separate classes fusion	分类融合
separating combining	分离-合并
sequence image segmentation	图像空间序列
sequence monocular image	单目序列影像
sequential adjustment	序贯平差
sequential approach	序贯法
sequential matching	串行匹配
sequential thinning	串行细化
serial charts	成套海图
series maps	系列地图
servo motor	伺服马达
set	集,测回
setting out of circular curves	圆曲线测设
setting out principal points	主要点测设

S

setting-out of main axis	主轴线测设
setting-out of reservoir flooded line	水库淹没线测设
setting-out survey	放样测量,施工放样
settlement	沉降
settlement observation	沉降观测,沉陷观测
settlement of buildings	建筑物沉陷
sewage system	污水系统
sextant	六分仪
shade	深色调
shadow analysis	阴影分析
shadow compensation	阴影补偿
shadow correction	阴影校正
shadow detection	阴影检测
shadow enhancement	阴影增强
shadow extraction	阴影提取
shadow removal	阴影去除
shaft	竖井
shaft orientation survey	竖井定向测量
shaft prospecting engineering survey	井探工程测量
shallow geological profile chart	浅地层地质剖面仪
shallowest sounding	最浅水深
shape	形状
shape features	形状特征
shape parameter	几何形状特征
shear	剪切力
shearing deformation	剪切变形
sheet designation, sheet number	图幅编号

S

sheet line system 地图分幅系统
sheet number 图号
sheet number system 图号系统
shield head 盾首
shield tunneling method 盾构开挖法
shift estimation 运动参数估计
ship noise 船舶噪声
ship position 船位
shoal 浅滩
short baseline acoustic system 短基线水声定位系统
shortest route 最短路径
short-range positioning system 近程定位系统
shot 镜头
shot boundary detection 镜头边缘检测
shrinkage 收缩
shrinkage and swell factors 收缩膨胀因子
side intersection 侧方交会
side look angel 扫描侧视角
side overlap, side lap 旁向重叠
side scan sonar 侧扫声呐
side-looking 侧视的
side-looking airborne radar (SLAR) 机载侧视雷达,侧视雷达
side-looking radar 侧视雷达
sidereal day 恒星日
sidereal time 恒星时
side-scanning mode 侧向扫描方式
sidetone 侧音

SIG	空间信息网格
sight line	视线
sighting axis	照准轴
sighting distance, sight length	视距
sighting line method	瞄直法
signal obstruction	信号障碍
signal-to-noise ratio	信噪比
signature	特征,标记,符号
significance level	显著性水平
significance test	显著性检验
significant figure	有效数字
silk-screen printing	丝网印刷
sill	海槛
similar figures	相似图形
similar triangles	相似三角形
similarity	相似性
similarity measure	相似性度量
simple Bouguer gravity anomaly	简单布格重力异常
simple damped harmonic motion	阻尼简谐运动
simple harmonic motion	简谐运动
simple iteration method	简单迭代法
Simpson's rule	森逊法则
simulated annealing arithmetic	模拟退火算法
simulated melt-annealing algorithm	模拟加温-退火算法
simulated remote sensing images	模拟遥感数据
simulation	模拟,模仿,仿真
simulation of ant colony behavior	蚁群行为仿真
simulation of atmospheric effects	大气辐射模拟

simulation of RS image	遥感图像模拟
simultaneous adjustment	整体平差
simultaneous contrast	视场对比
simultaneous equations	联立方程
simultaneous equations of planes	平面方程
simultaneous independent model aerotriangulation adjustment	独立模型法空中三角测量整体平差
simultaneous inequalities	联立不等式
simultaneous observation	同步观测
simultaneous, synchronous	同步的,同时的
sine	正弦
sine formula	正弦公式
SINEX	国际 GPS 数据处理结果的交换格式
single difference	单差
single difference phase observation	单差相位观测
single frame	单张像片
single frequency	单频
single point matching	单点匹配
single scattering albedo	单散射反照率
single sheet chart	单幅海图
single wire method: two shafts	两井定向
single-pass across track interferometry	单轨式横跨式轨迹干涉雷达测量系统
singleton	单元集
single-valued function	单值函数
singular	奇异的
singular matrix	奇异矩阵

S

singular value decomposition (SVD)	奇异值分解
singularity	奇异性,异常
sinker	沉锤
Sir-C	美国和加拿大发射的合成孔径测试雷达
site	工地,现场
site boundary	现场边界(线)
site plan, site map	工地(总)平面图
site preparation	场地平整
site reconnaissance	现场勘测
site selection	选址
skeleton	骨架化
skeleton drawing	草图,简图
skeleton line	骨架线
skeleton structure	框架结构
sketch	草图,概要
sketch design	设计草图
sketch drawing	草图,略图
skew distribution	偏斜分布
skew line	偏斜线
skewness	偏斜
SKI	Leica GPS 接收机随机软件
slack tide	平潮
slant height	斜高
slant range correction	倾斜距离校正
slave station	副台

S

Slichter modes(triplet) 核模
slide projector 滑动式投影仪
slip forming 滑模施工
slit 缝隙
slit light 缝光源
slope 坡度
slope current 倾斜流
slope distance 斜距
slope line 示坡线
SLR 卫星激光测距
small format aerial photography (SFAP) 小像幅航空摄影
small matrix array image 小面阵 CCD 影像
small scale 小比例尺
smallcraft chart, yacht chart 游艇用图
small-sample 小样本
smart media 智能介质卡
smart sensor web 智能传感器网格
smoothing 平滑
smoothing adjustment 使平滑,使光滑
smoothing techniques 平滑技术
smoothness 平滑度
snake model 活动轮廓
snow-cover extracting 雪盖提取
snowpack 雪盖
society control 社会控制
soft copy 软拷贝
soft thresholding filter 软门限滤波

softassign	软对应
softcopy photogrammetry	软拷贝摄影测量
software independent exchange (format)(SINEX)	国际GPS数据处理结果的交换格式
SOG	对地航速
soil erosion intensity	土壤侵蚀强度
soil moisture	土壤水分
soil salt	土壤含盐量
soil temperature	土壤温度
soil thermal inertia	土壤热惯量
soil water content	土壤含水量
Sokkia	日本索佳
solar calculation	日照计算
solar construction	太阳能建筑物
solar day	真太阳日
solar irradiance, solar radiation	太阳辐射
solar panel	太阳能板
solar parallax	太阳视差
solar radiation spectrum	太阳辐射波谱
solar spectrum	太阳光谱
solar system	太阳系
solar tide	太阳潮
solar time	真太阳时
solid Earth tide	固体潮
solid of revolution	旋转体
solid spherical harmonics	球体谐函数
solid state camera	固态摄像机
solid zonal harmonics	体带调和函数,体带谐函数

solid-state laser	固体激光器
solution	解
solution of equation	方程解
solution set	解集
sonar	声呐
sonar equations	声呐方程
sonar image	声呐图像
sonar image mosaic	声呐图像镶嵌
sonar sweeping	声呐扫海
sonobuoy	声呐浮标
SORA	奥地利 Vienna 大学研制的 DTM 软件
sound ray	声线
sound reflection	声反射
sound source	声源
sound speed profile	声速剖面
sound velocity correction	声速改正
sound velocity gradient	声速梯度
sound velocity in water	海水声速
sound velocity meter	声速仪
sound wave	声波
sound wave beam	声波束
sounder	声波测深器,发声器,测深仪
sounding	水深测量
sounding buoy	声学浮标
sounding datum	深度基准
sounding field book	水深测量手簿

S

sounding line	测深线
sounding point interval	水深点间隔
sounding range	测深范围
sounding rod	测深杆
source	信号源,辐射体
source coding	信源编码
source material, cartographic document	制图资料
south celestial pole	南天极
south ecliptic pole	南黄极
south hemisphere	南半球
south pole	南极
space design	空间设计
space geodesy	空间大地测量学
space geodetic techniques	空间大地测量技术
space intersection	空间前方交会
space microwave radiometer	星载微波辐射计
space model	空间模型,立体模型
space oblique azimuthal projection	空间斜方位投影
space photogrammetry	航天摄影测量,太空摄影测量
space photography	航天摄影
space planning	空间布局
space remote sensing	航天遥感
space resection	空间后方交会
space resource storage	空间资源储备
space science	空间科学
space segment	空间部分

space shuttle 航天飞机
space station 航天站
space structure 空间结构
space waves 空间波
space-borne 星载
spaceborne SAR 星载合成孔径雷达
spacecraft 宇宙飞船
spacial frequency 空间频谱
spacing 间隔,间距
spacing of buildings 建筑物间距
span 跨,跨度
sparseness 稀疏化处理
spatial analysis 空间分析
spatial association rule 空间关联规则
spatial attribute 空间属性
spatial autocorrelation 空间自相关
spatial clustering 空间聚类
spatial coordinates 空间坐标
spatial correlation 空间相关
spatial data 空间数据
spatial data clearing house 空间数据交换网站
spatial data infrastructure(SDI) 空间数据基础设施
spatial data mining 空间数据挖掘
spatial data structure 空间数据结构
spatial data transfer 空间数据转换
spatial data transfer standard (SDTS) 空间数据转换标准
spatial data warehouse 空间数据仓库

S

spatial database management 空间数据库管理
spatial database management system 空间数据库管理系统
spatial decision 空间决策
spatial distribution 空间分布
spatial distribution and temporal variation 时空分布
spatial feature 空间特征
spatial fuzzy membership 空间隶属度
spatial indexing 空间索引
spatial information grid (SIG) 空间信息网格
spatial information science 空间信息科学
spatial interpolation 空间插值
spatial modeling 空间建模
spatial object 空间对象
spatial orientation 空间定向
spatial query 空间查询
spatial rectangular coordinate 空间直角坐标
spatial reference system 空间参照系
spatial relationship 空间关系
spatial resolution 空间分辨率
spatial simulation 空间模拟
spatial topology 空间拓扑关系
spatial unit 空间单元
spatial-temporal data model 时空数据模型
spatiotemporal association rule 关联规则
spatio-temporal data 时空数据
spatio-temporal information 时变信息

special chart 专用海图
special committee for oceanographic research 海洋研究特别委员会
special economic zone 经济特区
Special International Committee on Antarctic Research 南极研究专设委员会
special relativity 狭义相对论
special road 专用道路
special use map 专用地图
specialized information grid 狭义空间信息网格
specification 规范,规程,施工说明书
specifications of survey 测量规范
speckle 斑纹,斑点
speckle noise 斑点噪声
speckle reduction 斑点噪声滤除
spectral 光谱的
spectral absorption 光谱吸收
spectral analysis 光谱分析,谱分析
spectral characteristic 光谱特性
spectral correlation 谱相关
spectral decomposition 光谱分解
spectral density function 谱密度函数
spectral distance 光谱距离
spectral indices 光谱指数
spectral information 波谱信息
spectral information characteristics 光谱信息特征
spectral matching 光谱匹配

spectral matching technique	光谱匹配技术
spectral measurement methods	光谱测量技术
spectral mixing	光谱混合,像元光谱混合
spectral preservation	光谱保持
spectral profile	光谱曲线
spectral range	光谱范围
spectral rebuilding	光谱重建
spectral response	光谱响应
spectral sensitivity	光谱感光度,光谱灵敏度
spectral shape	光谱形状
spectral signature	光谱特征
spectral simulation	光谱模拟
spectral variable	光谱变量
spectrograph	摄谱仪
spectrometer	波谱仪,频谱仪,光谱仪
spectroradiometer	分光辐射计
spectrum	光谱,频谱,波谱
spectrum cluster	波谱集群
spectrum dividing techniques	分光技术
spectrum structure	谱间结构
specular reflection	镜面反射
specular reflection	镜向反射
speed over ground (SOG)	对地航速
sphere	球形
spherical astronomy	球面天文学

S

spherical azimuth	椭球面方位角
spherical coordinate	球面坐标
spherical error probable	球概率误差
spherical harmonics	球谐函数
spherical triangle	球面三角形
spheroid	球体,椭球体
spheroid of revolution	旋转椭球
spin angular momentum	旋转角动量
spin axis	旋转轴
spin speed	旋转速度
spinning gyro	旋转陀螺
spiral	螺旋线
spiral curve geometry	螺旋线几何特性
spiral curve location	缓和曲线测设
spiral stairs	螺旋楼梯
spline method	样条函数法
split-window algorithm	劈窗算法
Spot 5	法国 Spotimage 公司的制图卫星系统,2002 年发射
SPOT satellite	SPOT 卫星
SPOT (System Probatoite d'observation de La Terre)	法国发射的用于地球资源遥感的卫星
spread function	扩散函数
spring block	拱座,弹簧座
spring tide	大潮
SPS	标准定位服务
spur leveling line	支水准路线

S

square	平方,正方形
square bracket	方括号
square control network	施工方格网
square map subdivision	正方形分幅
square matrix	方(矩)阵
square meter	平方米
square root	平方根
square symmetric matrix	对称方阵
squaremillimeter	平方毫米
SRROD filter	统计比值差值排序滤波器
SRTM(The Shuttle Radar Topography Mission)	微波遥感卫星,2000年发射
STA(star tracker)	恒星跟踪器
stability	稳定性
stability of structures	建筑物稳定性
stability test of echo sounder	测深仪稳定性检验
stabilized platform	稳定平台
stable anchor point	稳定锚固点
stable point	稳定点
stadia	视距,视距尺
stadia constant	视距常数
stadia hair	视距丝,视距线
stadia interval	视距间隔
stadia multiplication constant	视距乘常数
stadia reading	视距读数
stadia traverse	视距导线
staff	标尺

S

stake	桩,标桩
stake out	定线,放样
standard area of urban structure	标准型城市,标准结构城市
standard design	标准设计
standard deviation	标准差
standard dimension	标准尺寸
standard equation	标准方程
standard error	标准误差
standard field of length	长度标准检定场
standard hydrophone	标准水听器
standard meridian	标准子午线,标准经线
standard meter	线纹米尺,日内瓦尺
standard normal distribution	标准正态分布
standard of construction	施工标准,建筑标准
standard of digital map product	数字地图产品标准
standard parallel	标准纬线
standard positioning service (SPS)	标准定位服务
standard unit	标准单位
standard video format camera	标准视频(电视)幅面摄像机
standardised residual	标准残差
standardization	标准化
standpipe	竖管,管体式水塔
State Bureau of Surveying and Mapping (SBSM)	国家测绘局
static draft	静吃水

S

static load 静荷载
static positioning 静态定位
static sensor 静态传感器
static situation 静止状态
static theory of tides 潮汐静力学理论
station 测站
station centring 测站归心
station chain 台链
stationarity 平稳性
stationarity conditions 平稳性条件
stationary 平稳
stationary sea surface topography 稳态海面地形
stationary series 平稳序列
stationary stochastic model 平稳随机模型
stationary stochastic process 平稳随机过程
stationary value 平稳值
statistical analysis 统计分析
statistical data 统计数据
statistical fitting and stretching 统计拟合拉伸
statistical independence 统计独立
statistical map 统计地图
statistical significance 统计显著性
statistical testing 统计检验
statistics 统计学
steel tape 钢尺
steerable filter 方向可调滤波器
steering compass 操舵罗经
stein estimator 压缩估计

stellar camera	恒星摄影机
stellar parallax	恒星视差
steller calibration	恒星检校
step edge curve	刀刃曲线
step pulse	步进脉冲
step-by step design	分段设计
stereo	立体的,立体观测的,立体镜的
stereo compensation	立体补偿
stereo eyewear, stereo glasses	立体眼镜
stereo images	立体影像
stereo view	立体观察
stereo vision	双目立体视觉
stereoautograph	立体自动测图仪
stereoblock adjustment	立体区域网平差,三维区域网平差
stereocamera, stereometric camera	立体摄影机
stereocomparator	立体坐标量测仪
stereocompilation	立体测图
stereographic projection	球面投影
stereointerpretoscope	立体判读仪
stereometer	立体量测仪,视差量测仪,视差杆,体积计
stereometric	立体测量的
stereopair, stereo image pair, stereo photopair	立体像对
stereophotogrammetry	立体摄影测量
stereoplotter	立体测图仪

stereoplotting	立体绘图
stereosat	立体卫星
stereoscope	立体镜
stereoscopic	立体的,立体观测的
stereoscopic (vision) acuity	立体视觉敏感度
stereoscopic map	视觉立体地图
stereoscopic model	立体模型
stereoscopic observation	立体观测
stereoscopic vision	立体视觉
stereoscopically	立体地,体视地
stochastic model	随机模型
stochastic process	随机过程
stochastic properties of the observations	观测值的随机特性
Stokes theory	斯托克斯理论
stop watch	秒表
stop-and-go positioning	走走停停定位法
storm surge	风暴潮
straight line	直线
straight line detection	直线检测
straight line navigation	直线导航
straight-line extraction	直线抽取
straight-line traverse	直伸导线
strain	张力,应变
strain ellipse	应变椭圆
strain gauge	应变计,张力计
strain rosette	应变片花
strain tensor	应变张量

S

strait	海峡
strategy for urban redevelopment	城市发展战略
streaking noise, striping noise	条带噪声
stream direction	流向
strictly monotonic	严格单调
strictly monotonic function	严格单调函数
strip	扫描带,带状
strip aerial triangulation	航带法空中三角测量
strip map, strip plan	带状图
strip mosaic	航带拼接
structural element	结构元
structure algorithm	结构算法
structure analysis	结构分析
structured selection	结构化选取
strut	支柱，压杆
sub-pixel	亚像元
sub-bottom profiler	浅地层剖面仪
Sub-Committee on Safety of Navigation	海运安全组织
subdivisional organization	再分结构
submarine cable	海底电缆
submarine canyon	海底峡谷
submarine construction survey	海底施工测量
submarine control network	海底控制网
submarine geographic chart	海底地貌图
submarine pipeline	海底管道
submarine plateau	深海高原
submarine range	海底山脉

submarine relief	海底地势
submarine ridge	海岭
submarine situation chart	海底地势图
submarine structure chart	海底地质构造图
submarine tunnel survey	海底隧道测量
submarine volcano	海底火山
sub-mean filter algorithm	亚均值滤波
subpixel	子像素,亚像素
subpixel matching	子像素匹配
sub-satellite point	卫星星下点
sub-sequence	子(序)列
subset	子集
subsidence	沉降
subsidence observation	沉降观测,沉陷观测
subsidiary angle	辅助角
subspace	子空间
substructure	下部结构,基础,地下建筑
subtend	对向
subtidal zone	潮下带
subtraction	减法
subway survey	地下铁道测量
successive approximation	逐次逼近法
successive contrast	连续对比
successive orientation	连续定向
sufficiency	充分性
sufficient and necessary condition	充要条件
sufficient condition	充分条件

S

suffix	下标
summation	求和法
summation formula	求和公式
sun compass	太阳罗经
sun-synchronous satellite	太阳同步卫星
super dimension of spectral space	超维光谱空间
super highrise building	超高层建筑
super resolution	超分辨率
super set	母集
superconducting gravimeter	超导重力仪
superelevation	超高
SuperMap GIS	超图地理信息系统软件
superposition	叠合
super-short baseline acoustic system	超短基线水声定位系统
supersonic echo sounder	超声波回声测深仪
superstructure	上部结构,上层建筑
supervised classification	监督分类
supervising engineer	监理工程师
supplementary contour	间曲线,半距等高线,补充等高线
support system	支撑体系
support vector machine(SVM)	支撑向量机,支持向量机
support vector regression	支撑向量回归
surface albedo	地表反照率
surface control	地面控制

surface control network	地面控制网
surface control survey	地面控制测量
surface deformation	地表变形
surface of position(SOP)	位置面
surface reconstruction	表面重建
surface reflectance	地表反射率
surface relief	地势,地面起伏
surface soil moisture	地表层湿度
surface spherical harmonics	球面调和函数,球面谐函数
surface temperature	表面温度,地面温度
surface water capacity index(SWCI)	地表含水量指数
surface/underground connecting survey	地面/地下联系测量
survey adjustment	测量平差
survey bracket	测量支架(托架)
survey buoy	测量浮标
survey coordinate	测量坐标
survey crew	测量组
survey datum	测量基准
survey in mining pit	采区测量
survey line	测线
survey marker	测量标志
survey method	测量方法
survey of present state at industrial site	工厂现状图测量
survey robot	测量机器人

S

survey specifications, specifications of surveys	测量规范
survey vessel	测量船
surveying	测量学
surveying and mapping	测绘
surveying and mapping of Antarctica	南极制图
surveying control network	测量控制网
surveying for site selection	厂址测量
suspended gyro	悬挂式陀螺
suspended sediment concentration	悬浮泥沙
suspension bridge	悬索桥
suspension tape(wire)	悬带
sustainable development	可持续发展
sustainablility	可持续性
SVD	奇异值分解
SVM	支撑向量机，支持向量机
swath	卫星扫描带，条带
swath echo sounder	条带测深仪
swath width	扫描带宽
sweep area	扫海区
sweep blind zone	扫测盲区
sweep survey	扫海测量
sweeper	扫海具
sweeping	扫海
sweeping at definite depth	定深扫海
sweeping depth	扫海深度

S

sweeping with side scan sonar	侧扫声呐扫海
swell	膨胀,涌浪
swing adjustment	旋角校正
swing angle, yaw	像片旋角
symbol identification	标志识别
symbolization	符号化
symbols and abbreviations on chart	海图图式
symmetric point	对称点
symmetric relation	对称关系
symmetric wavelets	对称小波
symmetry	对称性
synchronous experiment	同步测量实验
syncline	[地]向斜
synthetic aperture	综合孔径
synthetic aperture radar(SAR)	合成孔径雷达
synthetic aperture sonar	合成孔径声呐
synthetic map	合成地图
system control center(SCC)	系统控制中心
system integration	系统集成
system of assessment	评价体系
systematic error	系统误差
systematic sampling	系统采样

T

tacheometry, stadia	视距测量(法)
tactual map, tactile map	触觉地图
TAI	国际原子时
tails off	拖尾
tangent distance	切距
tangent offset method	切线支距法
tangent point	切点
tangential acceleration	切向加速度
tangential distortion, tangential lens distortion	切向畸变
tangential parallax	切向视差
target	觇牌
target centring	照准点归心
target detection	目标检测
target location	目标定位
target of sonar	声呐目标
target reflector	目标反射器
targets extraction	目标提取
TASH	德国 Hannover 大学研制的 DTM 软件
tatget point	照准点
Taylor expansion	泰勒展开式
Taylor series	泰勒级数
TBM method	隧道开挖法

T

TDOP	时间精度因子
TDT	地球质心力学时
technical economic index	经济技术指标
technical report	技术报告
technical specification	技术规范
technique of city planning	城市规划技术
technogical term	技术术语
tectonic deformation	构造变形
tectonic motion	构造运动
tectonic phenomena	构造现象
telescope	望远镜
telluroid	近似地形面
temperature correction	温差改正
temperature difference	温差
temperature effect of gravimeter	重力仪的温度影响
temperature of sea water	海水温度
temperature profile	温度廓线
temperature sensor	温度传感器
temperature/salinity diagram	温-盐图解
template matching	模板匹配
temporal characteristic	时态特征
temporal differencing	时间差分
temporal resolution	时间分辨率
temporal variation	时变
temporary building	临时性建筑，临时性房屋
temporary construction	临时工程，临时建筑
temporary population	暂住人口

temporary works	临时工程
tension, tensional stress	张力
tensor	张量
term of life	使用期限
term of validity	有效期限
terminal box	终端框
terminal point	终点
terminal side	终边
terminal velocity	终端速度
terrain	地势, 地形, 地面
terrain analysis	地形分析
terrain change	地表形变
terrain characteristic	地形特征线
terrain feature	地形要素
terrain matching	地形匹配
terrain model	地形模型
terrain modeling	地形建模
terrain profile	地形断面
terrain telemetry	地形遥测
terrain undulation	地形起伏
terrain visualization	地形可视化
terrestrial camera	地面摄影机
terrestrial coordinate system	地面坐标系
terrestrial dynamic time (TDT)	地球质心力学时
terrestrial photogrammetry	地面摄影测量, 地形摄影测量
terrestrial photograph	地面摄影像片
terrestrial spectrograph	地面摄谱仪

T

terrific raid accident	恐怖袭击事件
territorial planning	国土规划
territorial seas	领海
territory	领域,领土
tertiary	第三位
tesseled cap transformation	穗帽变换
tesseral spherical harmonics	田球调和函数,田球谐函数
test criterion	检验标准
test of multple hypotheses	多维假设检验
test of significance	显著性检验
test range calibration	实验场检校
tetrahedron	四面体
texture	纹理
texture analysis	纹理分析
texture classification	纹理分类
texture compression	纹理压缩
texture enhancement	纹理增强
texture feature	纹理特征
texture image	纹理影像
texture image segmentation	纹理分割
texture index	纹理指数
texture mapping	纹理映射
texture restoration	纹理图像复原
texture spectrum	纹理谱
TGO	天宝 GPS 随机数据处理软件
Thales	法国天泰雷兹

the second moment 二阶矩
the specification for oceanographic survey 海洋调查规范
thematic atlas 专题地图集
thematic attribute 专题属性
thematic cartography 专题地图学
thematic chart 专题海图
thematic map 专题图
thematic mapper 专题测图仪
thematic mapping 专题制图
thematic overlap 专题层
theodolite 经纬仪
theorem 定理
theoretical cartography 理论地图学
theoretical foundation 理论基础
theoretical geodesy 理论大地测量学
theoretical lowest tide surface 理论最低潮位面
theoretical probability 理论概率
theory of errors 误差理论
theory of least squares 最小二乘理论
theory of probability 概率论
theory of surveying errors 测量误差理论
theory of testing hypotheses 假设检验理论
thermal agitation 热搅动,热激发
thermal imager 热像仪
thermal inertia 热惯量
thermal infrared 热红外
thermal infrared detector 热红外探测器

thermal infrared imagery, thermal IR imagery	热红外图像
thermal infrared multispectral scanner(TIMS)	热红外多光谱
thermal infrared remote sensing, thermal IR remote sensing	热红外遥感
thermal infrared scanner	热红外扫描仪
thermal radiation	热辐射
thermal source	热源
thermocline	温度跃层
thermograph	温度计
thinning algorithm	细化算法
third-order design (THOD) : the improvement problem	三类设计:网的改造
three dimensional coordinate	三维坐标,空间坐标,立体坐标
three line CCD scanners	三线阵 CCD
three line scanner	三线阵扫描仪
three-carrier ambiguity resolution (TCAR)	三频相位模糊度解算
three-dimensional city modeling	三维城市建模
three-dimensional display	三维显示
three-dimensional geodesy	三维大地测量学
three-dimensional information	三维信息
three-dimensional network	三维网
three-dimensional simulation	三维模拟
three-dimensional space	三维空间
three-dimensional terrain simulation	三维地景仿真

T

three-line array	三线阵
three-line array CCD imagery	三线阵 CCD 影像
three-wire leveling	三丝水准测量
threshold	门槛,临界值
threshold analysis	临界分析,门槛理论
threshold / thresholding value	阈值
tidal analysis	潮汐分析
tidal chart	潮汐图
tidal current	潮流
tidal current chart	潮流图
tidal datum	潮汐基准
tidal day	潮日
tidal factor	潮汐因子
tidal harmonic constant	潮汐调和常数
tidal information panel	潮信表
tidal light	潮汛灯
tidal observation	潮汐观测
tidal perturbation	潮汐摄动
tidal phenomena	潮汐现象
tidal prediction	潮汐预报
tidal quasi-harmonic analysis	潮汐准调和分析
tidal range	潮差
tidal rise	潮升
tidal staff	验潮水尺
tidal station	验潮站
tidal station level mark	验潮站水位标志
tidal theory	潮汐理论
tidal wave	潮汐波

tidal well	验潮井
tide	潮汐
tide constant	潮汐常数
tide correction	潮汐改正
tide gauge	验潮
tide height datum	潮高基准面
tide synobservation	同步验潮
tide table	潮汐表
tide-generating force	引潮力
tide-generating potential	引潮位
tie point	连接点
tilt adjustment	倾角改正
tilt angle of photograph	像片倾角
tilt observation	倾斜观测
time systems	时间系统
time dilution of precision(TDOP)	时间精度因子
time domain	时域
time invariant linear system	时不变线性系统
time of high water	高潮时
time of low water	低潮时
time series analysis	时间序列分析
time series graphs	时间序列图
time zone	时区
time-consuming	费时
time-domain descriptions of time series	时间序列时域描述
time-of-week(TOW)	周计数
time-varying coordinate	时变坐标

T

TIN	不规则三角网
tint	浅色调
tint graduation	分层设色表
tip	倾斜,倾角,尖端
tolerance	限差,容许误差
tolerance radius	限差范围
tomographic scanning	层析扫描
tone(response) adjustment	色调调整
Topcon	日本拓普康
TOPEX/POSEIDON(T/P)	托帕克斯卫星
topocentric cartesian coordinate	站心直角坐标
topocentric coordinate	站心坐标
topocentric coordinate system	站心坐标系
topocentric origin	站心原点
topocentric spherical coordinate	站心球面坐标
topographic base map	地形底图
topographic base mapping	地形地图制图,地形地图测图
topographic correction, terrain corrections	地形改正
topographic database	地形数据库
topographic indentification	地形判读
topographic map	地形图
topographic map of urban area	城市地形图
topographic map symbols	地形图图式
topographic mapping	地形制图,地形测图
topographic patterns	地形特征
topographic survey	地形测量

T

topography	地形学,地形测量学
topological analysis	拓扑分析
topological data model	拓扑数据模型
topological map	拓扑地图
topological relation	拓扑关系
topological retrieval	拓扑检索
topological structure	拓扑结构
topology	拓扑学,地志学
topometric data	地形测量数据
TopoMouse	三维量测鼠标
toponomastics, toponymy	地名学
torque eliminated	无扭
total accuracy of sounding	测深总精度
total correction of echo sounder	回声测深仪总改正
total electron content(TEC)	总电子含量
total linear closure of traverse	导线全长闭合差
total magnetic intensity chart	海洋全磁力图
total span	总跨度
total station	全站仪
tourist map	旅游地图
tow transducer	拖曳式换能器
tow vehicle	拖曳载体
tower	塔
towertop	塔顶
town and country planning	城乡规划
town center	城镇中心
town cluster	城镇群
town planning	城镇规划,城市规划

town planning map 城镇规划图
tracBack 按航迹返航
trace of a matrix 矩阵的迹
tracing 跟踪,蒙绘
tracing of check line 检查线透写图
tracing of sounding line 测深线透写图
track (TRK) 航向,航迹
track adjustment 航道校正,轨道调整
track estimation 航迹推算
track plotter 航迹自绘仪
tracker 跟踪仪
tracking accuracy 跟踪精度
tracking station 跟踪台
track-up display 航向向上显示
traditional architecture 传统建筑
traffic accident 交通事故
traffic assignment 交通分配
traffic lane 交通道
traffic volume 交通量
traffic way 车道,交通路线
training 训练
training samples 训练样本
transceiver 步话机,收发两用机
transducer 换能器
transducer dynamic draft 换能器动态吃水
transducer of sounder 测深仪换能器
transducer static draft 换能器静态吃水
transducer tow vehicle 换能器拖体

transfer format	交换格式
transfer function	传递函数
transfer of elevation	高程传递
transformation equation	转换方程
transit	经纬仪
transit axis	水平轴,横轴,旋转轴
transit instrument	经纬仪
transit method	中天法
transition curves	过渡曲线
transition layer of sound velocity	声速跃层
transitive property,transitivity	传递性
translation	平移
translation parameters	平移参数
transmission electron microscope	透射电子显微镜
transmission spectrum	透射谱
transmittance	透射,透光率
transmitting line of sounder	测深仪发射线
transparent foil	网纹片
transponder	转发器,脉冲转发机
transpose	移项,转置
transpose of matrix	转置矩阵
transverse axis	横轴
transverse coast	横向海岸
transverse component	横截分量
transverse Mercator chart	横轴墨卡托投影地图
transverse projection	横轴投影
transverse proportion distortion	横向比例畸变
transverse resolution	横向分辨率

T

trapezium	梯形
trapezoidal integration	梯形集合
trapezoidal rule	梯形法则
travel graph	行程图
traverse angle	导线折角
traverse leg	导线边
traverse network	导线网
traverse point	导线点
traversing, traverse survey	导线测量
tree diagram	树形图
tree rows and hedges	行树与篱笆
trench	海沟
triangle	三角形
triangle inequality	三角不等式
triangle method	三角形法
triangular matrix	三角矩阵
triangulated irregular network (TIN)	不规则三角网
triangulateration	边角测量
triangulateration network	边角网
triangulation	三角测量
triangulation base	三角测量基线
triangulation chain	三角锁
triangulation network	三角网
triangulation point	三角点
tribrach	三角基座
trifocal tensor	三焦张量
trigonometric control	三角测量控制

T

trigonometric equation	三角方程
trigonometric function	三角函数
trigonometric identity	三角恒等式
trigonometric leveling	三角高程测量
trigonometric leveling network	三角高程网
trigonometric ratio	三角比
trigonometric table	三角函数表
trigonometry	三角学
trilateration	三边测量
trilateration network	三边网
trim	纵倾
Trimble	美国天宝
trinomial	三项式
triple	三倍
triple angle	三倍角
triple difference phase observation	三差相位观测
triple product	三重积
triple-difference	三差
tripod	三脚架
tripod base	三脚架基座
trisect	三等分
troposphere	对流层
tropospheric delay	对流层延迟
tropospheric refraction correction	对流层折射改正
trough	海槽
true azimuth	真方位角
true color	真彩色
true course	真航向

T

true error	真误差
true horizon	真地平线,真水平线
true meridian	真子午线
true north	真北
true sidereal time	真恒星时
true solar day	真太阳日
true solar time	真太阳时
true value	真值
true vernal equinox	真春分点
truncation error	截断误差
truss	桁架,桁梁,构架
truss span	桁架跨度
trussed beam	桁架梁
tunable filter	可调谐滤光片
tunnel boring machine (TBM) method	隧道开挖法
tunnel breakthrough	隧道贯通
tunnel control station	隧道控制点
tunnel guidance system	隧道导向系统
tunnel profiler	隧道断面仪
tunnel shield machine	隧道盾构法
tunnel survey	隧道测量
TurboSurvey	ROUGE GPS 接收机随机软件
turning point	转点
turning point (or reversal) method	逆转点法
turning point plate	尺垫
turning radius	转弯半径

T

twist	扭曲
two-color laser ranger	双色激光测距仪
two-dimensional space	二维空间
two-medium photogrammetry	双介质摄影测量
two-sided test	双边检验
two-tail test	双尾检验
two-way route	双向航道
two-wire method：one shaft	一井定向
typal map	类型地图
type Ⅰ error	Ⅰ型误差
type Ⅱ error	Ⅱ型误差
typical construction	典型建筑，典型工程

T

U

U. S. Naval Oceanographic Office	美国海军海洋局
UAV(Unmanned Aerial Vehicle)	无人驾驶飞行器
UAVRSS(unmanned aerial vehicle remote sensing system)	无人机遥感系统
UGIS	城市地理信息系统
ultimate load	最大荷载
ultlasonic scanner	超声扫描仪
ultrasonic hologram system	水生全息摄影系统
ultrasonogram gradient change	声图灰度变化
ultrasonogram structure	声图结构
umbrella	测伞
unbiased estimate	无偏估计
unbiased estimator	无偏估计量
unbounded function	无界函数
uncertainty	不确定性
uncertainty analysis	不确定性分析
uncontrolled mosaic	无控制点镶嵌图,像片略图
uncorrelated observations	无关观测值
undecimated wavelet transform	无抽样小波变换
under color addition	底色增益
under color removal	底色去除
undercrossing	地下通道
underground alignment	地下定线

underground building	地下建筑
underground cavity survey	井下空硐测量
underground control	地下控制
underground control survey	地下控制测量
underground leveling	地下水准测量
underground mining	地下采矿
underground pipeline survey	地下管线测量
underground railway	地下铁路
underground survey	地下测量
underground traversing	地下导线测量
underground utilities	地下设施
undersampled	欠采样
underwater acoustics	水声学
underwater camera	水下摄影机
underwater laser range finder	水下激光测距仪
underwater photogrammetry	水下摄影测量
underwater sound communication station	水声通讯台
underwater stereoscopic photographic apparatus	水下立体摄影仪
underwater television	水下电视
undetermined coefficient	待定系数
unequal	不等
unequally weighted series	不等权序列
uneven grey rectification	灰度不均匀校正
ungrouped data	不分组数据
uniform	一致(的),均匀(的)
uniform acceleration	匀加速度

U

uniform color space	均匀颜色空间
uniform motion	匀速运动
uniform speed	匀速率
uniform velocity	匀速度
uniformly distributed	均匀分布
unimodal distribution	单峰分布
union entropy	联合熵
unique solution	唯一解
unique value	单值图
uniqueness	唯一性
unit area	单位面积
unit circle	单位圆
unit matrix	单位矩阵
unit vector	单位向量
unit volume	单位体积
unit weight	单位权
United States Geological Survey	美国地质测量局
Universal Polar Stereographic projection(UPS)	通用极球面投影
universal theodolite	全能经纬仪,通用经纬仪
universal time coordinated(UTC)	世界协调时
universal time(UT)	世界时
universal transverse Mercator projection(UTM)	通用横轴墨卡托投影
unknow quantity(parameter)	未知量(参数)
unknown	未知数
unknown position	未知点位

unmaned air vehicle for remote sensing system 无人机遥感监测系统
unsupervised classification 非监督分类
untrained types 未训练类别
unwrap 解缠
up neighbor run-length 上邻游程
update 更新
update rate 更新速率
updating and surveying 变更调查
upgrade 上坡
upland 山地,丘陵地
upland plain 高原
up-link station 注入站
upper bound 上界
upper limit 上限
upper transit 上中天
upper triangular matrix 上三角阵
uppertidal zone 潮上带
UPS 通用极球面投影
up-to-date map 现势地图
up-to-date style 现代风格
urban aesthetic 城市美学
urban analysis 城市分析
urban and regional planning 城市与区域规划
urban area 城市地区,市区
urban building 城市建筑
urban built-up area 城市建成区
urban comprehensive planning 城市综合规划

urban construction administration 城市建设管理
urban control survey 城市控制测量
urban crisis 城市危机
urban demography 城市人口学
urban design 城市设计
urban detailed planning 城市详细规划
urban development strategy 城市发展战略
urban disaster prevention 城市防灾
urban disasters 城市灾害
urban ecological system 城市生态系统
urban economy 城市经济
urban effect 城市效应
urban emergency rescue 城市应急救援
urban flood control 城市洪水控制
urban framework 城市结构,城市格局
urban function 城市功能
urban fundamental geographic information 城市基础地理信息
urban geographic information 城市地理信息
urban geographic information system(UGIS) 城市地理信息系统
urban gridding 城市网格化
urban heat island 城市热岛
urban infrastructure 城市基础设施
urban infrastructure planning 城市基础设施规划
urban land 城市用地
urban management 城市管理
urban man-made hazards 城市人为灾害

urban mapping	城市制图
urban national disasters	城市自然灾害
urban network	城市网络
urban overall planning	城市总体规划
urban planning	城市规划
urban planning area	城市规划区
urban planning land use administration	城市规划用地管理
urban population	城市人口[居民]
urban problem	城市问题
urban public transportation system	城市公共交通系统
urban redevelopment	城市改建
urban remote sensing technology	城市遥感技术
urban safty	城市安全
urban scale	城市规模
urban scale of air pollution	城市空气污染标度
urban sewerage and drainage	城市排水
urban sudden public health accidents	城市突发公共卫生事件
urban survey	城市测量
urban sustainable development	城市可持续发展
urban system	城市体系
urban system planning	城镇体系规划
urban technologic disasters	城市技术灾害
urban thematic geographic information	城市专题地理信息
urban transportation	城市交通运输，市区运输

urban trend	城市化趋向
urban water management	城市用水管理
urbanization	城市化,都市化
urbanization level	城市化水平
urbanology	城市学,城市问题研究
urban-rural population	城乡人口
usable area	使用面积
user accuracy	用户精度
user equivalent range error	用户等效距离误差
user segment	用户部分
UT0(universal time-zero)	世界时
UT1(universal time-one)	世界时(加上极移改正)
UT2(universal time-two)	平世界时
UTC	世界协调时
UTM	通用横轴墨卡托投影
u-v coverage	u-v 覆盖

V

vacant land	闲置地,空地
validity	有效性
valley line	山谷线
vandalism	故意破坏行为
vanishing point	合点,灭点
vanishing point control	合点控制
variability	可变性
variable	变量
variable speed	可变速率
variable velocity	可变速度
variance	方差
variance factor	方差因子
variance inflation model	方差膨胀模型
variance of unit weight	单位权方差,方差因子
variance within clusters	类内方差
variance-covariance matrix	方差-协方差矩阵
variance-covariance propagation law	方差-协方差传播律
variation	变数
varied information entropy	变率信息熵
variomat	变线仪
varioscale projection	变比例投影
vectograph method of stereoscopic viewing	偏振光立体观察

vector 向量
vector data 矢量数据
vector data structure 矢量数据结构
vector map 矢量地图
vector plotting 矢量绘图
vector quantization 矢量量化
vector radiative transfer theory (VRT) 矢量辐射传输理论
vector space 矢量空间,向量空间
vector tracking 矢量跟踪
vectorization 矢量化
vector-to-raster conversion 矢量-栅格转换
vegetated surface 植被覆盖
vegetation cover change 植被覆盖变化
vegetation index 植被指数
vegetation spectrum 植被光谱
vegetation visualization 植被可视化
vegular matrix 可逆矩阵
vehicle-based image sequence 车载序列影像
velocimeter 声速计
velocity hydrophone 阵速水听器
velocity made good (VMG) 沿计划航线上的航速
velocity of wave group 波群速度
Vening-Meinesz formula 维宁曼尼斯公式
venomous chemical accident 有毒化学品灾害
verhicle compass 车载罗盘
vernal equinox 春分或秋分,昼夜平分点

vernier accuracy	游标精度
vertical acceleration effect	垂直加速度影响
vertical accuracy	高程精度
vertical aerial photograph	垂直航空摄影像片
vertical alignment	竖向定线
vertical angle	垂直角
vertical asymptote	垂直渐近线
vertical axis	竖轴,纵轴
vertical bubble	垂直水准器
vertical circle	垂直度盘
vertical clearance	竖向净空
vertical component	垂直分量
vertical control densification	高程控制加密
vertical control network	高程控制网
vertical control point, vertical control station	高程控制点
vertical control survey	高程控制测量
vertical coordinates	垂直坐标
vertical curve	竖曲线
vertical curve geometry	竖曲线几何形状
vertical curve location	竖曲线测设
vertical design	竖向设计
vertical displacement	垂直位移
vertical displacement observation	垂直位移观测
vertical distance	垂直距离
vertical distribution	垂直分布
vertical epipolar line	垂核线
vertical epipolar plane	垂核面

V

vertical gradient of gravity 重力垂直梯度
vertical hair 垂直丝
vertical line 铅垂线,垂线
vertical line locus(VLL) 铅垂线轨迹法
vertical looking radar 垂直波束雷达
vertical parallax,y-parallax 上下视差
vertical photogrammetry 正直摄影测量
vertical photography 竖直摄影
vertical refraction coefficient 垂直折光系数
vertical refraction error 垂直折光差
vertical resolution 垂直分辨率
vertical settlement 竖向沉降
vertical shaft 竖井
vertical survey 高程测量,垂直测量
verticality control 铅直度控制
verticle circle 垂直度盘,竖盘
very long baseline interferometry (VLBI) 甚长基线干涉测量
vibration spectrum 振动频谱
vicarious external calibration 外定标
video compression 视频压缩
video object 视频对象
video photogrammetry 视频摄影测量
view 视图
view-frustum culling 视区裁剪
viewing angles 视角
village in the city 城中村
virtual city 虚拟城市

V

virtual landscape	虚拟地景
virtual map	虚地图
virtual real-2D field	虚拟真二维控制场
virtual reality technology	虚拟现实技术
virtual reality(VR)	虚拟现实
virtual reference station(VRS)	虚拟参考站
VirtuoZo	中国 SuperSoft 公司的卫星遥感测图处理系统
visibility acuity	能见敏锐度
visibility function	可视度函数
visible and near-infrared(VNIR)	可见近红外
visual analysis	视觉分析
visual balance	视觉平衡
visual contrast	视觉对比
visual field	视野
visual hierarchy	视觉层次
visual information	视觉信息,直观信息
visual interpretation	目视判读
visual landscape	视觉景观
visual measurement	视觉测量
visual tracking	视觉跟踪
visual variable	视觉变量
visual zenith telescope	目视天顶仪
visualization	可视化
visualization interface	可视化接口
visualization of spatial information	空间信息可视化
VLBI	甚长基线干涉测量

volcanic island	火山岛
volcano eruptions	火山爆发
volume calculation	体积计算
volume rendering	体视化
voxel	体素
VR	虚拟现实
VRS	虚拟参考站
V-STARS	瑞士 Leica 公司和美国 GSI 公司联合推出的视频摄影测量系统

W

WADGPS	广域差分 GPS
walking-talking	步话机,收发两用机
walkthrough	漫游
Walsh transformation	沃尔什变换
WAP GIS	无线网地理信息系统
water supply engineering	给水工程,供水工程
water body	水体
water contamination	水污染
water density chart	海水密度图
water drainage works	排水工程
water extraction	抽取水
water identification	水体识别
water level	水位
water pressure measurement	水压测量
water salinity chart	海水盐度图
water sampler	采水器
water spectrum	水体光谱
water supply volume	用水量
water surface	水面
water system	水系
water temperature chart	海水温度图
water transparency chart	海水透明度图
water vapor radiometer	水汽辐射仪
water vapor retrieval	水汽反演

waterlog damage	渍害
watershed area	分水区,汇水区域
watershed planning	流域规划
watershed transform	分水岭变换
wave amplitude	波幅
wave beam angle	波束角
wave buoy	波浪浮标
wave characteristic	波浪要素
wave crest	波峰
wave crest line	波峰线
wave current	波流
wave forecasting	海浪预报
wave forecasting chart	海浪预报图
wave gauge	测波仪
wave group	波群
wave height	波高
wave length	波长
wave loop	波腹
wave node	波节
wave observation	波浪观测
wave period	波周期
wave steepness	波陡
wave trough	波谷
waveform	波形
waveform analysis	波形分析
waveform retracking	波形重定
wavelength	波长
wavelet analysis	小波分析

wavelet coefficient standard deviation	小波系数标准方差
wavelet histogram	小波直方图
wavelet interpolation	小波插值
wavelet packet	小波包
wavelet transform	小波变换
wavelet transform method	小波变换法
wavelet-packet transform	小波包变换
waypoint	航路点
Waypoint	加拿大诺瓦泰公司 GPS 数据后处理软件
weak pixel	病态像元
weather disaster	气象灾害
web GIS	万维网地理信息系统
weight	权
weight coefficient	权系数
weight function	权函数
weight fusion	加权融合
weight matrix	权矩阵
weighted mean value, weighted average value	加权平均值
Weisbach triangle	韦史巴赫三角形
weiss quadrilateral	韦斯四边形法
wetland	湿地
WGS-84	1984 世界大地坐标系
white noise	白噪声
wide area augmentation system (WAAS)	广域增强系统

W

wide area differential GPS (WADGPS)	广域差分 GPS
wide band	宽波段
wide-angle digital camera	宽角数字相机
wide-band multi-spectral space	宽带多光谱空间
widelane	宽巷
wind retrieval	风场反演
wind-driven current	风海流
windowing	开窗
Winer spectrum	维纳频谱
wire drag survey	扫海测量
wireless aplication protocol GIS (WAP GIS)	无线网地理信息系统
workflow	工作流程
works limit	施工范围
working drawing	施工图
workstation	工作站
world atlas	世界地图集
world geodetic system 1984(WGS-84)	1984 世界大地坐标系
world oceanic chart	世界大洋全图
wriggle survey	收方测量

W

X

x-axis	x 轴
x-coordinate or northing	x 坐标
x(y) direction displacement	x(y)方向位移
xerography	静电复印
Xi'an geodetic coordinate system 1980	1980 西安坐标系
x-intercept	x 轴截距
XML	可扩展置标语言
x-parallax	左右视差

Y

y-axis	y 轴
Y-code, encrypted code	加密码
y-coordinate or easting	y 坐标
y-intercept	y 轴截距

Y

Z

zenith	天顶
zenith distance, zenith angle	天顶距
zero adjustment	归零,零位调整,零点调整
zero baseline	零基线
zero calibration	零点校正
zero check	零点检查
zero contour	零米等深线
zero drift	零点漂移
zero factor	零因子
zero intermediate frequency (ZIF) vector filtering	零中频矢量滤波
zero matrix	零矩阵
zero meridian	首子午线
zero point of tide staff	水尺零点
zero signal of echo sounder	测深仪零位信号
zero vector	零向量
zero-order design (ZOD): the datum problem	零类设计:基准设计
zero-order model	零阶模型
zig-zag traverse	Z 形导线
zonal harmonic coefficient	带谐系数
zonal rectification	分带纠正
zonal spherical harmonics	球带调和函数,球带谐函数

zone control	区域控制
zone dividing meridian	分带子午线
zone plate	波带板
zoom	缩放
zoom in	缩小
zoom out	放大
Z-tracking technique	Z 跟踪技术

Z

汉 英

数字·英文

1956 黄海高程系统	Huanghai vertical datum of 1956
1980 国际大地测量参考系统	geodetic reference system 1980 (GRS1980)
1980 国际大地测量参考系统	GRS1980
1980 西安坐标系	Xi'an geodetic coordinate system 1980
1984 世界大地坐标系	WGS-84
1984 世界大地坐标系	world geodetic system 1984 (WGS-84)
1985 国家高程基准	national height datum 1985
Ashtech GPS 接收机随机软件	GPPS, WinPRISM, Solution
BP 神经网络模型	BP ANN
CCD 面阵扫描器	CCD Array Scanners
CCD 线阵扫描器	CCD Line Scanners
CCD 摄影机	CCD camera (charge-coupled device camera)
CCD 推扫影像	CCD push-broom image
CCD 影像	CCD imagery
Geosurv 公司 GPS 数据处理软件	Flykin
GLONASS 接收机	GLONASS receiver
GPS 标准数据交换格式	RINEX

GPS 测量	GPS surveying
GPS 接收机	GPS receiver
GPS 空中三角测量	GPS aerotriangulation
GPS 时间	GPS time (GPST)
GPS 时间	GPST
GPS 信号频率之一(1227.6 MHz)	L2 frequency
GPS 信号频率之一(1575.42 MHz)	L1 frequency
GPS 星座	GPS constellation
GPS 仪器公司	GARMIN
GPS 载波相位	GPS carrier phase
GPS 周	GPS week
G-逆	g-inverse
II 型误差	type II error
Intergraph 公司推出的微机版三维计算机辅助设计系统	MicroStation
I 型误差	type I error
JAVAD GPS 接收机随机软件	Pinnacle
K-均值算法	K-means algorithm
L&R 海洋重力仪	Lacoste-Romberg marine gravimeter
Leica GPS 接收机随机软件	LGO,SKI
Leica 经销的机载数字传感器(三线阵数码相机)	ADS40

Leica 经销的数字摄影测量系统	Helava
L 波段	L-band
Moore-Penrose 广义逆	Moore-Penrose inverse
P 码,精码	P-Code
ROUGE GPS 接收机随机软件	TurboSurvey
SAR 复数图像	SAR complex image
SPOT 卫星	SPOT satellite
UNB 大学 GPS 数据处理软件	KARS
u-v 覆盖	u-v coverage
x(y)方向位移	x(y) direction displacement
x 轴	x-axis
x 轴截距	x-intercept
x 坐标	x-coordinate or northing
y 轴	y-axis
y 轴截距	y-intercept
y 坐标	y-coordinate or easting
Zeiss 的解析测图仪 C100 附加一对 CCD 相机构成的混合数字摄影测量系统	Indu SURF (Industrial Surface Measurement)
Z 跟踪技术	Z-tracking technique
Z 形导线	zig-zag traverse

A

阿达马变换	Hadamard transformation
安全出口	safty exit
安全间距,安全距离	safe distance
安全净空	safe clearance
安全区,安全带	safty strip
安置不准	misalignment
安置误差	boresight
安装测量	installation survey
岸线测量	coast line survey
按航迹返航	tracBack
暗礁	reef
暗色调,深色调	dark tone
凹	concave
凹多边形	concave polygon
凹向上的	concave up
凹向下的	concave down
奥地利 Vienna 大学研制的 DTM 软件	SORA

B

八边形	octagon
八杈树	octree
八面体	octahedron
巴尔达数据探测	Baarda's data snooping
坝址勘察	dam site investigation
白噪声	white noise
百分率	percentage
百分之	percent
百米	hectometre
摆	pendulum
摆幅，振幅	amplitude of oscillation
摆杆型海洋重力仪	beam-type sea gravimeter
摆线	cycloid
摆仪	pendulum instrument
斑点，小块地	patch
斑点噪声	speckle noise
斑点噪声滤除	speckle reduction
斑纹，斑点	speckle
板块构造学	plate tectonics
半参数回归	semi-parametric regression
半干旱草场	semiarid landscape
半共轭轴	semiconjugate axis
半径	radius, radii
半控制镶嵌图	semicontrolled mosaic

B

半球,半球体	hemisphere
半日潮	semidiurnal tide
半日潮港	semidiurnal tide harbor
半日潮流	semidiurnal tidal current
半色调	halftone
半圆	semicircle
半自动提取	semiautomatic extraction
伴随矩阵	adjoint matrix
饱和度	saturation
报警系统	alarm system
报警装置	alarm device
爆破(作业)	blasting
北	northing
北斗(卫星)	Beidou
北方向	north direction
北黄极	north ecliptic pole
北极	north pole
北极圈	arctic circle
北极星任意时角法	method by hour angle of Polaris
北天极	north celestial pole
北纬	northern latitude
贝塞尔大地主题解算公式	Bessel formula for solution of geodetic problem
贝塞尔椭球	Bessel ellipsoid, Bessel spheroid
贝叶斯定理	Bayes'theorem
贝叶斯方法	Bayesian approach
贝叶斯分类	Bayesian classification
贝叶斯估计	Bayesian estimation

贝叶斯网络	Bayesian Network(BN)
备份	backup
备选方案,(线路)比较方案	alternate plan
备选假设	alternative hypothesis
备选线路	alternative route
背景估计	background estimation
背景与目标	background and target
背景噪声	background noise
背斜	anticline
被动声呐	passive sonar
被动式传感器	passive sensor
被动式定位系统	passive positioning system
被动式遥感	passive remote sensing
被动微波	passive microwave
被动微波遥感	passive microwave remote sensing
被动微波遥感传感器	passive microwave sensor
被积函数	integrant
本初子午线	prime meridian
本征根,特征根	latent root
本征向量,特征向量	eigenvector
本征值,特征值	eigenvalue
比	ratio
比对点	comparison point
比辐射率封闭测定法	sealing cavity method
比较地图学	comparative cartography
比例	proportion
比例尺	proportional scale

比例尺,尺度　scale
比例尺变化　scale variation
比例尺变形　scale distortion
比例尺测定　determination of scale
比例尺常数　scaling constant
比例尺精度,尺度精度　scale accuracy
比例量表　ratio scaling
比例误差　proportional error
比例因数　scale factor
比特　bit
比特平面编码　bit-plane coding
比特-失真率　rate-distortion
比值变换　ratio transformation
比值分析　ratio analysis
比值增强　ratio enhancement
必然事件　certain event
必要条件　necessary condition
闭合差　closing error, closure, closure error, error of closure, misclosure
闭合导线　closed traverse
闭合环导线　closed loop traverse
闭合环线　closed loop
闭合曲线　closed curve
闭合水准路线　closed leveling line
闭合条件　condition of closure
闭区间　closed interval
闭凸区域　closed convex region

边长中误差	mean square error of length
边交会法	linear intersection
边角测量	triangulateration
边角交会法	linear-angular intersection
边角网	triangulateration network
边界	boundary
边界测量	boundary survey
边界调整	boundary adjustment
边界条件	boundary condition
边坡稳定性观测	observation of slope stability
边线,图廓线	border line
边线匹配	geographic line matching
边缘	edge
边缘保护	edge keeping
边缘点去除	boundary pixels removal
边缘海	marginal sea
边缘检测	edge detection
边缘连接	edge linking
边缘算子	edge operator
边缘提取	edge extraction
边缘增强	edge enhancement
编绘	compilation
编绘员	compiler
编码	encoding
编码度盘	circular encoder
编码法	coding method
编码数据	coded data
编译码	coding and decoding

编制图	compiled map
便携声呐	portable sonar
便携式回声测深仪	portable echo sounder
便携式验潮仪	portable tide gauge
变比例投影	varioscale projection
变更调查	updating and surveying
变化检测	change detection
变量	variable
变率	rate of change
变率信息熵	varied information entropy
变坡点	grade change point
变数	variation
变线仪	variomat
变形,畸变	deformation
变形参数	deformation parameter
变形分析	deformation analysis
变形改正	correction for deformation
变形观测控制网	control network for deformation observation, deformation observation control network
变形机理	mechanisms and physics of the deformation
变形监测(观测)	deformation monitoring (observation)
变形解释	deformation interpretation
变形体	deformable body
变形椭圆	indicatrix
变形向量	deformation vector

变形预测 prediction of deformation
标称精度 nominal accuracy
标尺 rod, staff
标定基线 calibration baselines
标定矩阵 calibration matrix
标识码 identifier
标书 bid
标志识别 symbol identification
标准残差 standardised residual
标准差 standard deviation
标准尺寸 standard dimension
标准单位 standard unit
标准定位服务 standard positioning service (SPS)
标准方程 standard equation
标准化 standardization
标准配置点 gruber point
标准设计 standard design
标准视频(电视)幅面摄像机 standard video format camera
标准水听器 standard hydrophone
标准纬线 standard parallel
标准误差 standard error
标准型 normalized form
标准型城市,标准结构城市 standard area of urban structure
标准正态分布 standard normal distribution
标准子午线,标准经线 standard meridian

B

表观反射率	apparent reflectance
表面温度,地面温度	surface temperature
表面重建	surface reconstruction
冰川运动	glacier movement
冰河学	glaciology
冰后回弹	post-glacial rebound
冰雪覆盖区	ice-snow covered area
饼图	pie chart
并行处理	parallel processing
并行模板	parallel masks
并行算法	parallel algorithm
并行通道接收机	parallel channel receiver
病态问题	ill-posed problem
病态像元	weak pixel
波瓣	lobe
波长	wave length
波带板	zone plate
波陡	wave steepness
波峰	wave crest
波峰线	wave crest line
波幅	wave amplitude
波腹	wave loop
波高	wave height
波谷	wave trough
波节	wave node
波浪补偿	heave compensation
波浪补偿器	heave compensator
波浪浮标	wave buoy

波浪观测	wave observation
波浪要素	wave characteristic
波流	wave current
波罗-科普原理	Porro-Koppe principle
波谱集群	spectrum cluster
波谱信息	spectral information
波谱仪,频谱仪,光谱仪	spectrometer
波群	wave group
波群速度	velocity of wave group
波束角	wave beam angle
波束宽度	beam width
波纹条纹	moiré fringe
波形	waveform
波形分析	waveform analysis
波形重定	waveform retracking
波周期	wave period
玻尔兹曼常数	Boltzmann's constant
伯努利分布	Bernoulli distribution
伯努利试验	Bernoulli experiment
泊松分布	Poisson distribution
薄泥浆填塞	grouting
薄云检测	haze detection
薄云去除	haze removal
补偿大地水准面	compensated geoid
补偿流	compensation current
补偿器	compensator
补偿器补偿误差	compensation error

B

不变矩	invariant moment
不变性	invariance
不成比例的	disproportional
不等	unequal
不等号	inequality sign
不等权序列	unequally weighted series
不等式	inequality
不定积分	indefinite integral
不定积分法	idenfinite integration
不动产测量,权属测量	property survey
不分组数据	ungrouped data
不共线	non-collinear
不规则	irregular
不规则沉降	non-uniform settlement
不规则离散点	irregular points
不规则三角网	triangulated irregular network (TIN)
不可能事件	impossible event
不可用	outage
不可约性	irreducibility
不连续点	discontinuous point
不连续性	discontinuity
不确定,不明确	indeterminacy
不确定性	uncertainty
不确定性分析	uncertainty analysis
不稳定性	instability
不显著的	insignificant

不相交的	disjoint
不相交的集	disjoint sets
不一致性	inconsistency
布格改正	Bouguer correction
布格异常	Bouguer anomaly
布隆斯公式	Bruns formula
布隆斯椭球	Brun's spheroid
布耶哈马问题	Bjerhammar problem
步测,定步	pacing
步话机,收发两用机	transceiver, walking-talking
步进脉冲	step pulse
步行街	pedestian street
部分和	partial sum

C

采泥器	mud snapper
采区测量	survey in mining pit
采区联系测量	connection survey in mining pit
采水器	water sampler
采样	sample
采样,抽样	sampling
采样间隔	sampling interval
采样率	sampling rate
采样密度	sampling density
采油	oil extraction
彩色编码	color coding
彩色变换	color transformation
彩色复制	color reproduction
彩色航空影像	color aerial image
彩色合成仪	additive color viewer
彩色红外片,假彩色片	color infrared film,false color film
彩色片	color film
彩色摄影	color photography
彩色校样	color proof
彩色样图	color manuscript
彩色增强	color enhancement
彩色坐标系	color coordinate system
参差的,不统一的	heterogeneous

参考标志	reference mark
参考点	reference point
参考基站	reference base stations
参考基准	reference datum
参考基准影像	reference image
参考空间	reference space
参考框架	frame of reference
参考平面	reference plane
参考球体,参考椭球体	reference spheroid
参考数据	reference data
参考椭球	reference ellipsoid
参考椭球定位	orientation of reference ellipsoid
参考站	reference station
参考站接收机	reference receiver
参数,参变量	parameter
参数方程	parametric equation
参数模型	parametric model
参数配置	optimum configuration
参数平差,间接平差	parametric adjustment
残差	residual, residue
残差平方和	residual sum of squares
舱外天线	outboard antenna
操舵罗经	steering compass
操作,运算	operation
草地	grass land
草图	draft plan
草图,概要	sketch

草图,简图	skeleton drawing
草图,略图	sketch drawing
侧方交会	side intersection
侧扫声呐扫海	sweeping with side scan sonar
侧扫声呐	side scan sonar
侧视的	side-looking
侧视雷达	side-looking radar
侧向扫描方式	side-scanning mode
侧音	sidetone
测标	measuring mark
测波仪	wave gauge
测风仪,风速风向仪	anemometer
测杆	measuring bar
测高	altimetric measurement
测高法,高程测量法	altimetry
测高仪	altimeter
测回	observation set
测绘	surveying and mapping
测绘车	mobile mapping vehicle
测绘工程	geomatics engineering
测绘工程师	geomatics engineer
测绘学	geomatics
测绘仪器	instrument of geomatics engineering
测角精度	accuracy of angular measurement
测角精度	angular accuracy
测角中误差	mean square error of angle observation
测距	distance measurement

C

测距定位系统 range positioning system
测距精度 accuracy of ranging, distance accuracy
测距雷达 range-only radar
测距误差 distance-measuring error
测距仪 distance measuring instrument, rangefinder
测控条 control strip
测量 measurement
测量标志 survey marker
测量船 survey vessel
测量等级 order of survey
测量方法 survey method
测量浮标 survey buoy
测量规范 specifications of survey, survey specifications, specifications of surveys
测量机器人 geo-robot, survey robot
测量基准 survey datum
测量控制网 surveying control network
测量频率 measurement frequency
测量平差 adjustment of observations, survey adjustment
测量误差理论 method of parameter eatimations, theory of surveying errors
测量校正 metric calibration
测量学 surveying
测量支架(托架) survey bracket

测量组	survey crew
测量坐标	survey coordinate
测流浮标	current measurement buoy
测伞	umbrella
测深	bathymetry
测深侧扫声呐	bathymetric sidescan sonar
测深定线	alignment of sounding
测深范围	sounding range
测深杆	sounding rod
测深归算	reduction of sounding
测深基准面	datum of soundings
测深密度	frequency of sounding
测深索	lead line
测深线	sounding line
测深线间隔	interval of survey line
测深线透写图	tracing of sounding line
测深仪记录器	recorder of soundings
测深仪	depth sounder
测深仪读数精度	reading accuracy of sounder
测深仪发射线	transmitting line of sounder
测深仪改正数	correction of sounder
测深仪换能器	transducer of sounder
测深仪回波信号	echo signal of sounder
测深仪记录	echogram
测深仪记录纸	recording paper of sounder
测深仪零位信号	zero signal of echo sounder
测深仪器	depth instrument
测深仪设计声速	designed sound velocity of sounder

C

测深仪稳定性检验	stability test of echo sounder
测深总精度	total accuracy of sounding
测图,制图	mapping
测图航空摄影	aerophototopography
测图卫星	Mapsat
测微密度计	microdensitometer
测线	survey line
测斜仪	inclinometer and tiltmeter
测站	station
测站归心	station centring
测站子午圈	observer's meridian
层次结构	level of detail
层间改正	plate correction
层析扫描	tomographic scanning
插值	interpolation
插值多项式	interpolating polynomial
插值法	method of interpolation
插值基函数	interpolation base function
插座,插孔	power outlet
差	difference
差分 GPS	differential GPS (DGPS)
差分定位	differential positioning
差分方程	difference equation
差分改正	differential correction
差分干涉测量法	differential interferometry
差分合成孔径雷达干涉测量	DInSAR
觇牌	target

C

长(度)	length
长半轴	semimajor axis
长除法	long division method
长度标准检定场	standard field of length
长方体	rectangular block
长方形	rectangle
长基线水声定位系统	long baseline acoustic system
长距离导航系统	long-range navigation system
长期磁变	magnetic secular change
长周期自由振荡	long-period free oscillation
常差	constant error
常规法	conventional method
常平架	gimbal suspension
常数	constant
常数项	constant term
常微分方程	ordinary differential equation
常用对数	common logarithm
常用坐标系,惯用坐标系	conventional coordinate system
常住人口	permanent population
嫦娥卫星	Chang-E satellite
厂房	factory building
厂址测量	surveying for site selection
场地,工地	lot
场地平整	site preparation
超导重力仪	superconducting gravimeter
超短基线水声定位系统	super-short baseline acoustic system

超分辨率	super resolution
超高	superelevation
超高层建筑	super highrise building
超光谱	hyperspectrum
超光谱成像仪	hyperspectral imager
超几何分布	hypergeometric distribution
超焦点距离	hyperfocal distance
超近景摄影测量	macrophotogrammetry
超媒体	hypermedia
超平面分割	hyperplane segmentation
超谱图像	hyperspectral image
超声波回声测深仪	supersonic echo sounder
超声扫描仪	ultlasonic scanner
超图	hypergraph
超图地理信息系统软件	SuperMap GIS
超维光谱空间	super dimension of spectral space
潮差	tidal range
潮高基准面	tide height datum
潮间带	intertidal zone
潮流	tidal current
潮流图	tidal current chart
潮日	tidal day
潮上带	uppertidal zone
潮升	tidal rise
潮汐	tide
潮汐表	tide table
潮汐波	tidal wave

潮汐常数	tide constant
潮汐调和常数	tidal harmonic constant
潮汐调和分析	harmonic analysis of tide
潮汐动力论	dynamical theory of tides
潮汐非调和常数	nonharmonic constant of tide
潮汐非调和分析	nonharmonic analysis of tide
潮汐分析	tidal analysis
潮汐改正	tide correction
潮汐观测	tidal observation
潮汐基准	tidal datum
潮汐静力学理论	static theory of tides
潮汐理论	tidal theory
潮汐摄动	tidal perturbation
潮汐图	tidal chart
潮汐现象	tidal phenomena
潮汐因子	tidal factor
潮汐预报	tidal prediction
潮汐准调和分析	tidal quasi-harmonic analysis
潮下带	subtidal zone
潮信表	tidal information panel
潮汛灯	tidal light
车道,交通路线	traffic way
车载罗盘	verhicle compass
车载序列影像	vehicle-based image sequence
沉锤	sinker
沉降	settlement, subsidence
沉降观测,沉陷观测	settlement observation, subsidence observation

C

衬砌	lining
成本	cost
成比例	proportional
成分分析	component analysis
成套海图	serial charts
成图比例尺,规定比例尺	proper scale
成像雷达	imaging radar
成像	image
成像干涉光谱仪	imaging Fourier transform spectrometer
成像光谱仪	imaging spectrometer
成像链	image chains (IC)
成像模型	imaging model
成像系统	imaging system
成像仪	imager
承重,承载能力	load bearing
承重能力	lifting capacity
承重墙	load bearing wall
城际交通枢纽	inter-city transportation
城际铁路	interurban railroad
城市安全	urban safty
城市测量	city survey, urban survey
城市测量数据库	database for urban survey
城市道路	municipal road
城市地理信息	urban geographic information
城市地理信息系统	urban geographic information system(UGIS)

C

城市地区,市区	urban area
城市地形图	topographic map of urban area
城市发展目标	goal for urban redevelopment
城市发展战略	strategy for urban redevelopment, urban development strategy
城市防灾	urban disaster prevention
城市分类	city classification
城市分析	urban analysis
城市改建	urban redevelopment
城市公共交通系统	urban public transportation system
城市功能	urban function
城市供热	municipal heating
城市供水	municipal water suply
城市管理	urban management
城市规划	city planning, municipal planning, urban planning
城市规划,城市布局	city layout
城市规划法规	legislation on urban planning
城市规划纲要	city planning outline
城市规划管理	city planning administration
城市规划技术	technique of city planning
城市规划区	urban planning area
城市规划用地管理	urban planning land use administration
城市规模	city size, urban scale
城市洪水控制	urban flood control
城市化	citify
城市化,都市化	urbanization

C

城市化趋向	urban trend
城市化水平	urbanization level
城市基础地理信息	urban fundamental geographic information
城市基础设施	urban infrastructure
城市基础设施规划	urban infrastructure planning
城市技术灾害	urban technologic disasters
城市建成区	urban built-up area
城市建设管理	urban construction administration
城市建筑	municipal architecture, urban building
城市交通运输，市区运输	urban transportation
城市结构	city structure
城市结构,城市格局	urban framework
城市经济	urban economy
城市景观	civic landscape
城市可持续发展	urban sustainable development
城市空气污染标度	urban scale of air pollution
城市控制测量	urban control survey
城市扩建	municipal extension
城市扩张,城市膨胀	city expansion
城市美学	urban aesthetic
城市排水	municipal drainage, urban sewerage and drainage
城市平面图	city plan
城市群	city agglomeration
城市热岛	urban heat island

C

城市人口[居民]	urban population
城市人口学	urban demography
城市人为灾害	urban man-made hazards
城市设计	urban design
城市生态系统	urban ecological system
城市体系	urban system
城市突发公共卫生事件	urban sudden public health accidents
城市网格化	urban gridding
城市网络	urban network
城市危机	urban crisis
城市问题	urban problem
城市详细规划	city detailed planning, urban detailed planning
城市效应	urban effect
城市性质	designated function of city
城市选址	city sitting
城市学,城市问题研究	urbanology
城市遥感技术	urban remote sensing technology
城市应急救援	urban emergency rescue
城市应急联动系统	city emergency response system (CERS)
城市用地	urban land
城市用水管理	urban water management
城市与区域规划	urban and regional planning
城市灾害	urban disasters
城市整体布局	overall urban layout

城市职能	city function
城市制图	urban mapping
城市中心区	inner city
城市专题地理信息	urban thematic geographic information
城市自然灾害	urban national disasters
城市综合规划	urban comprehensive planning
城市总体规划	urban overall planning
城市总体规划纲要	master planning outline
城乡规划	town and country planning
城乡人口	urban-rural population
城镇规划,城市规划	town planning
城镇规划图	town planning map
城镇群	town cluster
城镇体系规划	urban system planning
城镇中心	town center
城中村	village in the city
乘	multiply
乘常数	multiplication constant
乘法	multiplication
乘积,积	product
乘数,乘式	multiplier
程控仪器	program-controlled instrument
吃水,草图,设计图样	draft
吃水改正	draft correction
迟延，滞后	lag
尺垫	turning point plate
尺度参数	scale parameter

C

尺度共生矩阵	scale based concurrent matrix
尺度间相关模型	interscale model
尺度空间	scale space
尺度空间表达	scale-space representation
尺度内相关模型	intrascale model
尺度效应	scaling effect
赤道	equator
赤道半径	equatorial radius
赤道长度比	equatorial scale
赤道面	equatorial plane
赤道圈	equatorial circle
赤道视差	equatorial parallax
赤道重力	equatorial gravity
充电电池	rechargeable battery
充分条件	sufficient condition
充分性	sufficiency
充要条件	sufficient and necessary condition
抽取水	water extraction
抽象符号	abstract symbol
抽象数据类型	abstract data type
抽样定理	sampling theorem
抽样分布	sampling distribution
出版原图	final original
出水口,电源插座	outlet
初版海图	new chart
初步设计图	preliminary drawings
初测,勘测	preliminary survey, reconnaissance survey

初等函数	elementary function
初等矩阵	elementary matrix
初始定位时间	acquisition time
初始化	initialization
初始近似值	initial approximation
初速度	initial velocity
初值,始值	initial value
初值问题	initial-value problem
除法	division
除法算式	division algorithm
处理框	process box
触觉地图	tactual map, tactile map
传递函数	transfer function
传递性	transitive property, transitivity
传感器姿态	sensor attitude
传送带	coveyer belt
传统大地测量监测技术	classical geodetic monitoring techniques
传统建筑	traditional architecture
船舶噪声	ship noise
船位	ship position
船位推算	dead reckoning
串行匹配	sequential matching
串行细化	sequential thinning
垂核面	vertical epipolar plane
垂核线	vertical epipolar line
垂球	plumb bob
垂线	vertical line

C

垂线,垂直(于)	perpendicular
垂线方向	direction of plumb-line
垂线偏差	deflection of the vertical
垂线偏差改正	correction for deflection of the vertical
垂心	orthocentre
垂直波束雷达	vertical looking radar
垂直的,正态的,正常的	normal
垂直度盘,竖盘	vertical circle, elevation circle, verticle circle
垂直分辨率	vertical resolution
垂直分布	vertical distribution
垂直分量	vertical component
垂直航空摄影像片	vertical aerial photograph
垂直加速度影响	vertical acceleration effect
垂直渐近线	vertical asymptote
垂直角	vertical angle
垂直距离	vertical distance
垂直平分线	perpendicular bisector
垂直水准器	vertical bubble
垂直丝	vertical hair
垂直位移	vertical displacement
垂直位移观测	vertical displacement observation
垂直折光差	vertical refraction error
垂直折光系数	vertical refraction coefficient
垂直坐标	vertical coordinates
垂准仪,铅垂仪	plumb aligner

垂足	foot of perpendicular
春分点	point of Aries, vernal equinox
春分或秋分,昼夜平分点	vernal equinox
磁北	magnetic north
磁北极	north magnetic pole
磁测深仪	magnetic sounder
磁测站	geomagnetic survey station
磁差图	magnetic variation chart
磁场强度	magnetic field strength
磁定向	magnetic orientation
磁方位角	magnetic azimuth
磁航向	magnetic course
磁极	magnetic pole
磁计程仪	electromagnetic log, EM log
磁力梯度仪	magnetic gradiometer
磁力线	magnetic line of force
磁力仪,地磁仪	magnetometer
磁力异常	magnetic anomaly
磁力异常区	magnetic anomaly area
磁罗经	magnetic compass
磁盘阵列	disk array
磁偏常数	declination constant
磁偏点	declination station
磁偏角	compass declination, magnetic declination, magnetic declinetion
磁偏角弧	declination arc
磁倾角	magnetic inclination

C

磁纬度	magnetic latitude
磁象限角	magnetic bearing
磁异常剖面图	magnetic anomaly profile
磁针	magnetized needle
磁子午线	magnetic meridian
从起点到当前位置的方位	course made good (CMG)
从整体到局部	from the whole to the part
粗糙集理论	rough set theory
粗糙逻辑	rough logic
粗差检测	gross error detection
粗差探测	detection of outliers
粗调	coarse adjustment
粗集	rough set
粗略定向	preliminary orientation
粗码,C/A 码	Coarse/Acquision Code (C/A Code)
粗扫	coarse sweeping, preliminary sweep
错角	alternate angle

C

D

大坝变形观测	dam deformation observation
大坝监测	dam monitoring
大坝施工测量	dam construction survey
大比例尺	large scale
大比例尺地形图	large scale topographical map
大测距,远测距	far range
大潮	spring tide
大潮平均高潮位	mean high water springs
大地测量,大地测量学	geodetic surveying
大地测量边值问题	geodetic boundary value problem
大地测量参考系统	geodetic reference system
大地测量控制	geodetic control
大地测量控制网优化设计	optimal design of geodetic networks
大地测量平面基准	horizontal geodetic datum
大地测量数据库	geodetic database
大地测量学	geodesy
大地测量学家	geodetist
大地测量仪器	geodetic instrument
大地方位角	geodetic azimuth
大地高	ellipsoidal height,geodetic height
大地基准	geodetic datum
大地基准的天文大地定位	astro-geodetic datum orientation

D

大地经度	geodetic longitude
大地水准测量	geodetic leveling
大地水准面	geoid
大地水准面高,大地水准面差距	geoid height,geoid undulation
大地天顶延迟	atmospheric zenith delay
大地天文学	geodetic astronomy
大地网	geodetic network
大地纬度	geodetic latitude
大地纬圈	geodetic parallel
大地位置	geodetic position
大地线	geodesic
大地原点	geodetic origin
大地重力测量学	geodetic gravimetry
大地主题反解	inverse solution of geodetic problem
大地主题正解	direct solution of geodetic problem
大地坐标	geodetic coordinate
大地坐标系	geodetic coordinate system
大光斑激光雷达	large footprint lidar
大角度倾斜	high oblique
大陆边缘	continental margin
大陆岛	continental island
大陆架	continental shelf
大陆架地形测量	continental shelf topographic survey
大陆架地形图	continental shelf bathymetric chart
大陆隆	continental rise
大陆坡	continental slope

D

大气层	atmosphere
大气传输特性	characteristics atmospheric transmissivity
大气窗口	atmospheric window
大气点扩散函数	atmospheric point spread function
大气辐射传输	atmospheric radiative transfer
大气辐射传输方程	atmosphere transfer model
大气辐射模拟	simulation of atmospheric effects
大气改正,气象改正	atmospheric correction
大气科学	atmospheric science
大气濛雾	atmospheric fog
大气色散	atmospheric dispersion
大气水汽	atmospheric water
大气透过率	atmospheric transmissivity
大气吸收	atmospheric absorption
大气下行辐射效应	effect of atmospheric downward thermal radiance
大气消光	atmospheric extinction
大气效应	atmospheric artifact
大气压	atmospheric pressure
大气影响	atmospheric effects
大气影响校正	correction of atmospheric effects
大气噪声	atmospheric noise
大气折射	atmospheric refraction
大气阻力摄动	atmospheric drag perturbation
大圈,大圆	great circle
大数定理	law of large number
大像幅摄影机	large format camera(LFC)

D

大洋地势图	general bathymetric chart of the oceans
大洋地形图	ocean bathymetric topography
大洋图	ocean chart
大洋中脊	mid-oceanic ridge
大样本	large-sample
大圆航程	great circle distance
大圆航线图	great circle sailing chart
代换法,换元法	method of substitution
代价函数	cost function
带通滤波	band-pass filtering
带通滤波器	band-pass filter
带谐系数	zonal harmonic coefficient
带状图	strip map, strip plan
待定系数	undetermined coefficient
单边检验	one-sided test
单差	single difference
单差相位观测	single difference phase observation
单点定位	point positioning
单点匹配	single point matching
单调	monotone
单调递减	monotonicaly decreasing
单调递减函数	monotonicaly decreasing function
单调递增	monotonicaly increasing
单调递增函数	monotonicaly increasing function
单调函数	monotonic function
单调收敛性	monotonic convergence
单峰分布	unimodal distribution

单幅海图	single sheet chart
单轨式横跨式轨迹干涉雷达测量系统	single-pass across track interferometry
单目序列影像	sequence monocular image
单频	single frequency
单散射反照率	single scattering albedo
单色	monochrome
单位矩阵	identity matrix, unit matrix
单位面积	unit area
单位权	unit weight
单位权方差,方差因子	variance of unit weight
单位体积	unit volume
单位向量	unit vector
单位圆	unit circle
单项式	monomial
单像计算机视觉	monocular image computer vision
单要素分类	individual element classification
单元集	singleton
单张像片	single frame
单值函数	single-valued function
单值图	unique value
弹道摄影测量	ballistic photogrammetry
弹道摄影机	ballistic camera
当地平均海面	local mean sea level
挡土墙,护土墙	earth-retaining wall
刀刃曲线	step edge curve
导标	leading beacon

D

导电	electric conduction
导航	navigation
导航电文	navigation message
导航数据格式	NMEA
导航台	navigation station
导航台定位测量	navigation station location survey
导航图	navigation chart
导航卫星测时测距	NAVSTAR
导航线,叠标线	leading line
导数	derivative
导线闭合差	closure error of traverse, misclosure of traverse
导线边	traverse leg
导线测量	traversing, traverse survey
导线长度误差	linear error of traverse
导线点	traverse point
导线横向误差	lateral error of traverse
导线角度闭合差	angle closing error of traverse, angular closure of traverse
导线节点	junction point of traverse
导线平差	adjustment of traverse
导线全长闭合差	total linear closure of traverse
导线网	traverse network
导线相对闭合差	relative linear error of traverse
导线折角	traverse angle
导线纵向误差	longitudinal error of traverse
岛弧	island arc
岛架	insular shelf

D

岛间海	inter-island sea
岛链	island chain
岛屿	island
岛屿测量	island survey
岛屿联测	island-mainland connection survey
岛屿图	island chart
倒锤[线]观测,倒锤法	inverse plummet observation, inverted pendulum
倒数	reciprocal
到目的地的最佳行驶方向	course to steer(CTS)
道路(铁路)中线	centreline of the road or railway
道路净空	road clearance
道路搜索	road searching
道路探测/检测	road detection
道路提取	road extraction
道路网	road network
德国、瑞士、苏黎世合作研制的实时摄影测量系统	DIPS
德国 Hannover 大学研制的 DTM 软件	TASH
德国 Munich 大学研制的 DTM 软件	HIFI(the Height Interpolation by Finite Elements)
德国 Stuttgart 大学研制的 DTM 软件	SCOP
德国地学研究中心研发的精密定位与定轨软件	EPOS

D

灯标	light beacon
灯船	light ship
灯浮标	light buoy
灯光节奏	flashing rhythm of light
灯光射程	light range
灯光周期	light period
灯色	light color
灯塔	lighthouse
灯质	light characteristic
等(式)	equality
等比定理	equal ratios theorem
等比线	isometric parallel
等边(的)	equilateral
等边多边形	equilateral polygon
等边三角形	equilateral triangle
等变形线	distortion isograms
等差级数	arithmetic series
等磁力线	isodynamic line
等概率	equal probability
等高	isometry
等高距	contour interval
等高面	contour plane
等高线,轮廓线	contour
等高线标注	contour label
等高线跟踪	contour tracing
等高线绘制	contour drafting, contour drawing
等高线精度	contour accuracy
等高线图	contour map

等积三角形 equivalent triangle
等积投影 authalic projection, equal area projection, equal-area projection, equivalent projection
等积投影经度 authalic longitude
等积投影纬度 authalic latitude, equal latitude
等级结构 hierarchical organization
等集 equal sets
等价 equivalence
等价关系 equivalence relation
等价类 equivalence class
等角航法 rhumb line sailing
等角条件,正形投影 conformal projection
等角投影 equal-angle projection
等角投影经度,正形投影经度 conformal longitude
等距(的) equidistant
等距量表 interval scaling
等距投影 equidistant projection
等距圆弧格网 equilong circle arc grid
等量纬度 isometric latitude
等偏摄影 parallel-averted photography
等倾摄影 equally tilted photography
等深线 isobath
等视距 equal sighting lengths
等位面 equipotential surface
等温线 isotherm
等效黑体温度,等价黑体温度 equivalent black-body temperature

D

等效时间采样　equivalent time sampling
等效影像　equivalent image
等盐线　isosalinity line
等腰三角形　isosceles triangle
等值灰度尺　equal value gray scale
等值区域图　choroplethic map
等值线　isoline
等值线地图　isoline map
等值线法　isoline method
低岸线　low coastline
低层建筑　low rise building
低潮,低水位　low water
低潮高　height of low water
低潮时　time of low water
低潮线　low water line
低低潮　lower low water
低高潮　lower high water
低轨　low earth orbit (LEO)
低轨卫星　low earth orbit satellite (LEOS)
低空　low altitude
低水位基准面　low water datum
低通滤波　low-pass filtering
低通滤波器　low-pass filter
低纬度　low latitude
堤防,筑堤　embankment
堤基　embankment foundation
笛卡儿积　Cartesian product
笛卡儿平面　Cartesian plane

笛卡儿坐标系	Cartesian coordinate system
笛卡儿坐标	Cartesian coordiantes
底版测点	floor station
底层	lower layer
底点纬度	latitude of pedal
底角	base angle
底色去除	under color removal
底色增益	under color addition
底索	bottom wire
底质	bottom characteristics
底质采样	bottom characteristics sampling
底质调查	bottom characteristics exploration
底质分布图	bottom sediment chart
底质取样器	bottom sampler
地表变形	surface deformation
地表层湿度	surface soil moisture
地表反射率	surface reflectance
地表反照率	surface albedo
地表含水量指数	surface water capacity index (SW-CI)
地表实测数据	land-ground measurement data
地表温度反演	land surface temperature retrieval
地表形变	terrain change
地表移动观测站	observation station of surface movement
地磁北极	north geomagetic pole
地磁测量	magnetometry
地磁场	geomagnetic field

地磁极	geomagnetic pole
地磁记录仪	magnetograph
地磁台	magnetic station
地磁要素	magnetic elements
地磁要素梯度	magnetic element gradient
地磁子午线	geomagnetic meridian
地底点	ground nadir point
地点,位置	location
地方标准时	local standard time
地方城市	local city
地方恒星时	local sidereal time
地方基准,局部基准	local datum
地方平恒星时	local mean sidereal time(LMST)
地方时	local time
地方视时	local apparent time
地方太阳时	local solar time
地方天文时	local astronomical time
地方真恒星时	local apparent sidereal time (LAST)
地方真时	local true time
地方子午线	local meridian
地固坐标系	body-fixed coordinate system, earth-fixed coordinate system
地基系统	ground-based system
地基增强系统	ground-based augmentation system
地极	geographical pole
地籍	cadastre
地籍簿	land register

D

地籍测量	cadastral survey
地籍调查	cadastral inventory
地籍更新	renewal of the cadastre
地籍管理	cadastral survey management
地籍图	cadastral map, cadastral plan
地籍信息系统	cadastral information system
地籍修测	cadastral revision
地籍要素	cadastral feature
地籍制图	cadastral mapping
地价	land cost
地界	land boundary
地界测量	land boundary survey
地壳变形	crustal deformation
地壳均衡	isostasy
地壳均衡改正	isostatic correction
地壳形变观测	crust deformation measurement
地壳运动	crustal movement
地壳运动观测网	crustal movement observation network
地块测量	parcel survey
地垒	horst
地类界图	land boundary map
地理[坐标]参照	georeference
地理北极	north geographical pole
地理编码	geocoding
地理编码系统	geocoding system
地理标识符	geographic identifier
地理参考系	geographical reference system

D

地理方位角 geographical azimuth
地理格网 geographic grid
地理经度 geographic longitude
地理实体 geographic entity
地理视距 geographical viewing distance
地理数据 geographic data
地理数据库 geographic database
地理纬度 geographic latitude
地理纬圈 geographical parallel
地理位置 geographic position
地理相关模型 geo-relational model
地理信息 geographic information
地理信息传输 geographic information communication
地理信息服务 geographic information service
地理信息科学 geographic information science
地理信息系统 geographic information system (GIS)
地理信息元数据 metadata for geographic information
地理学,地理 geography
地理要素 geographic elements
地理置标语言 geographic markup language (GML)
地理中心 geographic center
地理资源分析支持系统 GRASS
地理坐标 geographic coordinates, geographic graticule

地幔　mantle
地貌图　geomorphological map
地貌形态示量图　morphometric map
地面/地下联系测量　surface/underground connecting survey
地面沉降　ground subsidence(settlement)
地面服务站　ground support system
地面高程　ground elevation
地面接收站　ground receiving station
地面控制　surface control
地面控制部分　ground-based control system (GCS)
地面控制测量　surface control survey
地面控制点　ground control point(GCP)
地面控制网　surface control network
地面摄谱仪　terrestrial spectrograph
地面摄影测量　terrestrial photogrammetry
地面摄影机　terrestrial camera
地面摄影像片　terrestrial photograph
地面实况　ground truth
地面照度　illuminance of ground
地面坐标系　terrestrial coordinate system
地名　geographic name, place name
地名录　gazetteer
地名数据库　geographic name database, place-name database
地名索引　geographic name index
地名通名　geographic general name

地名学	toponomastics, toponymy
地名正名	orthography of geographical name
地名转写	geographical name transcription, geographical name transliteration
地平面	horizon
地平视差,左右视差,横视差	horizontal parallax
地平线摄影机	horizon camera
地平线像片	horizon photograph
地坪	floor-on-grade
地堑	graben
地倾斜观测	ground tilt observation
地球摆动	Earth wobbles
地球扁率	Earth's flattening, flattening of the Earth
地球潮汐,固体潮	Earth tide
地球潮汐参数	Earth tidal parameters
地球定向参数	Earth orientation parameter(EOP)
地球动力扁率	dynamic ellipticity of the earth
地球动力因子	dynamic factor of the earth
地球固体内核	Earth's solid inner core
地球静止轨道卫星	geostationary orbit (satellite)
地球科学	Earth sciences
地球空间信息科学	geoinformatics
地球曲率	curvature of earth
地球岁差进动	Earth's precession
地球同步卫星	geo-synchronous satellite
地球椭球	Earth ellipsoid, Earth spheroid

地球位,大地位	geopotential
地球位数	geopotential number
地球物理测井	geophysical log
地球物理磁强计	geophysical magnetometer
地球物理地震法	seismic geophysical method
地球物理地震探测	geophysical seismic exploration
地球物理方法	geophysical method
地球物理勘测	geoexploration, geophysical exploration, geophysical prospecting
地球物理勘测仪	geophysical instrument
地球物理勘测仪器	exploration instrument for geophysics
地球物理卫星	geophysical satellite
地球物理效应	geophysical effect
地球物理学	geophysics
地球物理学家	geophysist
地球形状	Earth shape, figure of the Earth
地球旋转	Earth's spin
地球仪	globe
地球章动	Earth's (forced) nutation
地球振荡	Earth's oscillation
地球质心力学时	terrestrial dynamic time(TDT)
地球重力场	gravity field of the Earth
地球重力场模型	Earth gravity model
地球周年视运动(绕太阳旋转)	Earth's annual motion (revolution about the sun)
地球周日视运动(自转)	Earth's diurnal motion (spin)

D

地球资源卫星　Earth resources technology satellite (ERTS)
地球自由核章动　Earth's free core nutation(FCN)
地球自转　Earth's rotation
地球自转参数　Earth's rotation parameters(ERP)
地球自转角速度　rotational angular velocity of the earth
地区界限　area boundary
地势,地面起伏　surface relief
地势,地形,地面　terrain
地势图　hypsometric map
地速　ground speed
地图　map
地图比例尺　map scale
地图编绘　map compilation
地图编辑　map editing
地图编辑大纲　map editorial policy
地图编辑软件　cartographic editing software
地图变形　map distortion
地图表示法　cartographic presentation
地图查询　map query
地图尺寸测定　determination of map size
地图传输　cartographic communication
地图地物　cartographic feature
地图叠置分析　map overlay analysis
地图分幅系统　sheet line system
地图分类　cartographic classification
地图分析　cartographic analysis, map analysis

地图符号　map symbols
地图符号库　map symbol base
地图符号学　cartographic semiology
地图负载量　map load
地图复杂性　map complexity
地图复制　map reproduction
地图感受　map perception
地图格网　map grid
地图更新　map revision
地图规格　map specification
地图绘制,绘图　map plotting
地图集　atlas
地图集类型　atlas type
地图集信息系统　altas information system
地图简化　map generalization
地图精度　map accuracy
地图库　map library
地图利用　map use
地图量算法　cartometry
地图模型,制图模型　cartographic model
地图目录　map catalog
地图内容结构　cartographic organization
地图判读　map interpretation
地图拼接　map join
地图评价　cartographic evaluation
地图潜信息　cartographic potential information
地图清晰性　map clarity
地图色标　color chart, map color standard

D

地图色谱	map color atlas
地图设计	map design
地图收集	map collection
地图数据结构	map data structure
地图数据库	cartographic database
地图数据库管理系统	map database management system
地图数字化	map digitizing
地图投影	map projection
地图投影变形	map projection distortion
地图投影分类	map projection classfication
地图图层	map coverage
地图图幅	chart sheet
地图显示	map display
地图信息	cartographic information
地图信息系统	cartographic information system (CIS)
地图学	cartography,cartology
地图易读性	map legibility
地图印刷	map printing
地图语法	cartographic syntactics
地图语言	cartographic language
地图语义	cartographic semantics
地图语用	cartographic pragmatics
地图原图(包括实测、编绘、清绘原图)	map manuscript
地图阅读	map reading
地图整饰	map appearance
地图制图	map making

地图制图软件	cartographic software
地图注记	cartographic annotation
地图坐标原点	map origin
地物	ground feature
地物,地貌	land feature
地物波谱特性	object spectrum characteristics
地物符号,人工建筑符号	culture symbol
地下采矿	underground mining
地下测量	underground survey
地下导线测量	underground traversing
地下定线	underground alignment
地下管线测量	underground pipeline survey
地下建筑	underground building
地下控制	underground control
地下控制测量	underground control survey
地下设施	underground utilities
地下室	basement
地下水	ground water
地下水观测孔	ground water monitoring well
地下水情况	groundwater condition
地下水位	groundwater level
地下水准测量	underground leveling
地下铁道测量	subway survey
地下铁路	underground railway
地下停车场	base car park
地下通道	undercrossing
地心	centre of earth,geocenter

D

地心大地坐标	geocentric geodetic coordinate
地心地固	Earth centered Earth fixed
地心基准	geocentric datum
地心经度	geocentric longitude
地心纬度	geocentric latitude
地心向量径	geocentric radius vector
地心引力常数	geocentric gravitational constant
地心原点	geocentric origin
地心直角坐标系	geocentric rectangular coordinate system
地心坐标	geocentric coordinate
地心坐标系	geocentric coordinate system
地形,地貌,陆地轮廓	landform
地形测量	topographic survey
地形测量数据	topometric data
地形底图	topographic base map
地形地图制图,地形地图测图	topographic base mapping
地形断面	terrain profile
地形分析	terrain analysis
地形改正	topographic correction, terrain corrections
地形建模	terrain modeling
地形可视化	terrain visualization
地形模型	terrain model
地形判读	topographic indentification
地形匹配	terrain matching
地形起伏	terrain undulation

D

地形摄影测量	terrestrial photogrammetry
地形数据库	topographic database
地形特征	topographic patterns
地形特征,地形要素	relief feature
地形特征线	terrain characteristic
地形图	topographic map
地形图更新	revision of topographic map
地形图图式	topographic map symbols
地形学,地形测量学	topography
地形遥测	terrain telemetry
地形要素	terrain feature
地形制图,地形测图	topographic mapping
地学编码影像	geo-referenced image
地震	earthquake
地震波	seismic wave
地震回放仪	seismic signal processor
地震检波器	seismic detector
地震勘测法	seismic method
地震强度	earthquake intensity, seismic intensity
地震图	seismicity map
地震震级	earthquake magnitude, seismic magnitude
地震震源	earthquake origin
地址编码	address coding
地址地理编码	address geocoding
地址匹配	address matching
地质调查	geological survey

地质断层	geological fault
地质力学	geomechanics
地质裂缝	geological fissure
地质略图	geological scheme
地质剖面测量	geological profile survey
地质剖面图	geological section map
地质学	geology
地质灾害	geological hazard(disaster)
地质状况	geological condition
地轴	Earth axis
递减	decrease
递减函数	decreasing function
递减级数	decreasing series
递降序	descending order
递升序	ascending order
递推公式	recurrence formula
递增函数	increasing function
递增级数	increasing series
第二偏心率	second eccentricity
第二位,次级	secondary
第三位	tertiary
第一偏心率	first eccentricity
第一位,首级	primary
典型建筑,典型工程	typical construction
典型图形平差	adjustment of typical figures
典型相关	canonical correlation
点、线混合摄影测量	hybrid point-line photogrammetry
点方式	point mode

点估计	point estimate
点光源	point light source
点积	dot product
点扩散函数	point spread function(PSF)
点目标	point target
点位(坐标)计算	calculations of locations (coordinates)
点位中误差	mean square error of a point
点样本校验	point sample
点云	point cloud
点值法	dot method
点状符号	point symbol
电场	electric field
电磁波测距	electromagnetic distance measurement
电磁波测距仪	electromagnetic distance measuring instrument
电磁波频谱,电磁波谱,电磁光谱	electromagnetic spectrum
电磁波谱能	electromagnetic spectrum energy
电磁辐射	electromagnetic radiation
电导率传感器	conductivity sensor
电荷耦合器件	charge coupled device(CCD)
电荷注入器件	charge injection device(CID)
电极极化	electrode polarization
电离层	ionosphere
电离层延迟	ionospheric delay
电离层折射	ionospheric refraction

D

电离层折射改正	ionospheric refraction correction
电气竖井	electric shaft
电容水听器	capacitor hydrophone
电子测距仪	electronic distance measuring instrument(EDM)
电子出版系统	electronic publishing system(EPS)
电子传感器	electronic sensing device
电子导电体	electronic conductor
电子导航海图数据库	electronic navigation chart database (ENCDB)
电子地图	electronic map
电子地图集	electronic atlas
电子分色机	color scanner
电子海图	electronic chart
电子海图数据库	electronic chart database(ECDB)
电子海图显示和信息系统	electronic chart display and information system(ECDIS)
电子经纬仪	electronic theodolite
电子平板仪	electronic plane-table
电子求积仪	electronic planimeter
电子扫描	electronic scanning
电子扫描仪	electronic scanner
电子手簿,数据采集器	data recorder
电子水听器	electronic hydrophone
电子水准仪	electronic level
电子显微摄影测量	nanophotogrammetry
电子相关	electronic correlation

D

调绘	annotation
调焦	focus adjustment
调焦(检景器目镜的)	eyepiece adjustment
调焦误差	error of focusing, focusing error
调节器,传动装置,激励器,执行元件	actuator
调频频率	modulation frequency
调整[的]大地水准面	regularized geoid
调制	modulation
调制传递函数	modulation transfer function (MTF)
调制器	modulator
迭代	iteration
迭代法	iterative method
迭代反演	iterative inversion
迭代分解	iterative unmixing
迭代算法	iterative algorithm
迭代投影	iterative projection
迭代自组织数据分析算法	iterative self organizing data analysis techniques algorithm (ISODATA)
迭合	superposition
迭合法	method of superposition
叠加	overlay
叠置分析	overlay analysis
顶板测点	roof station
定标,定比例,量测	scaling
定轨	orbit determination

D

定积分	definite integral
定理	theorem
定量遥感	quantitative remote sensing
定量遥感模型	quantitative remote sensing model
定深扫海	sweeping at definite depth
定位	positioning, position fix
定位标志	positioning mark
定位点间距	positioning interval
定位与定向系统,POS 系统	position and orientation system (POS)
定线,放样	layout,stake out
定线,准直	alignment
定线测量	alignment survey
定线设计	alignment design
定向	orientation
定向参数	orientation parameters
定向测量	orientation survey
定向天线	directional antenna
定向运动地图	orienteering map
东经	east longitude
东距	easting
动吃水	dynamic draft
动荷载	dynamic load
动力(学)的	kinetic
动力大地测量学	dynamic geodesy
动力高改正	dynamic correction
动力海面地形	dynamic ocean topography
动力海洋学	dynamical oceanography

D

动力平衡,动态平衡	kinetic equilibrium
动力视差	dynamical parallax
动力学	dynamics
动力演化算法	dynamical evolutionary algorithm
动量	momentum
动圈换能器	moving-coil transducer
动态变化	dynamic change
动态变量	dynamic variable
动态传感器	dynamic sensor
动态窗口	dynamic window
动态的	kinematic
动态地景仿真	dynamic landscape simulation
动态地理信息系统	dynamic GIS
动态定位	kinematic positioning
动态规划	dynamic programming
动态监测	dynamic monitoring
动态系统	danamical system,dynamic system
动态预测	dynamic prediction
动态综合层次判别分析	dynamic hienchy discriminatory analysis
抖动	dithering
陡坡	abrupt slope
读数精度	reading accuracy
读数显微器	micrometer
独立分量分析	independent component analysis (ICA)
独立基线	independent baseline
独立模型法空中三角测量	independent model aerial triangulation

独立模型法空中三角测量整体平差 simultaneous independent model aerotriangulation adjustment
独立三角测量 independent triangulation
独立随机变量 independent random variable
独立网 isolated network
独立坐标系 independent coordinate system
度量关系 metric relation
度量空间 metric space
度量模型 metric model
度盘 circle
度盘分画 circle graduation
端点 end point
端元变化 selective endmember
端元自动提取 automated endmember extraction
短半轴 semiminor axis
短基线水声定位系统 short baseline acoustic system
短期磁偏差 magnetic declination variation
短轴 minor axis
断层,断裂 fault
断面测量 section survey
断面提取 profile extraction
对比度 contrast(CON)
对边 opposite side
对称点 symmetric point
对称方阵 square symmetric matrix
对称关系 symmetric relation
对称小波 symmetric wavelets
对称性 symmetry

对称轴 axis of symmetry
对地观测体系 Earth observing system (EOS)
对地观测卫星 Earth observing satellites
对地航速 speed over ground (SOG)
对地航向 course over ground (COG)
对光精度,调焦精度 focusing accuracy
对角 opposite angle
对角线 diagonal
对角线元素 diagonal element
对角阵 diagonal matrix
对流层 troposphere
对流层延迟 tropospheric delay
对流层折射改正 tropospheric refraction correction
对数 logarithm
对数方程 logarithmic equation
对数分布 logarithmic distribution
对数函数 logarithmic function
对向 subtend
对向观测 reciprocal observation
对向三角高程测量 reciprocal trig leveling
对象跟踪 object tracking
对中杆 centering rod
对中误差 centering error
盾构开挖法 shield tunneling method
盾首 shield head
钝角 obtuse angle
钝角三角形 obtuse angle triangle
多倍仪 multiplexer

多边形 polygon
多边形地图 polygonal map
多边形叠置 polygon overlay
多波段极化散射计 multi band polarimetric scatterometer
多波段遥感 multi-band remote sensing
多波束测深系统 multibeam bathymetric system
多波束测探 multibeam echosounding
多波束测探扫海 multibeam sounding sweeping
多波束测探系统 multibeam sounding system
多波束声呐 multibeam sonar
多层(建筑) multiple-story
多层建筑 multi-story building
多层结构 multilayer organization, multiple-structure
多车道高速公路 multilane highway
多尺度 multiscale
多尺度变换 multiscale transform
多尺度表达 multiscale representation
多尺度分割 multiresolution segmentation
多尺度分析 multiscale analysis
多尺度形态学 multiscale morphology
多传感器 multisensor
多次散射 multiple scattering, mutli-scattering
多点压力(压强)计 multi-point piezometer
多点影像匹配 multi-point matching
多分辨率 multiresolution
多分辨率分析 multiresolution analysis

D

多分辨率模型　multiresolution model
多光谱纹理　multispectral texture
多光谱遥感图像　multispectral remote sensed image
多光谱影像　multispectral image
多级纠正　multistage rectification
多焦点投影　polyfocal projection
多角度的　multi-directional
多角度光谱成像仪　multi-angle imaging spectroraiometer(MISR)
多结构元　multi-structure elements
多进制小波　M-band wavelet
多跨结构　multiple-span structure
多跨梁　multiple-span girder
多类　multi-class
多路复用接收机　multiplexing receiver
多路复用通道　multiplexing channel
多路径　multipath
多路径效应　multipath effect
多路误差　multipath error
多媒体地图　multimedia map
多面体　polyhedron
多年平均海面　multi-year mean sea level
多普勒测距系统　Doppler ranging
多普勒轨道定位与无线电集成系统　Doppler Orbit determination and Radiopositioning Integrated on Satellite (DORIS)
多普勒频移　Doppler shift
多普勒声呐　Doppler sonar

多谱段扫描仪	multispectral scanner (MSS)
多谱段摄影	multispectral photography
多谱段摄影机	multispectral camera
多谱段图像	multispectral imagery
多谱段遥感	multispectral remote sensing
多色海图	color chart
多摄站摄影测量	multistation photogrammetry
多时相	multitemporal
多时相 SAR 影像	multitemporal SAR image
多时相分析	multitemporal analysis
多时相遥感	multitemporal remote sensing
多时相影像	multitemporal images
多时相影像数据管理	multitemporal image management
多时相组合法	multitemporal composition
多天线 GPS 系统	multi-antenna GPS system
多通道	multichannel
多通道接收机	multichannel receiver
多通道滤波	multichannel filtering
多维假设检验	test of multple hypotheses
多维随机变量	multidimensional random variable
多线程	multithread
多相机组合	multi-camera system
多项式	polynomial
多项式方程	polynomial equation
多项式回归	polynomial regression
多项式平差	polynomial adjustment
多项式因式分解	factorization of polynomial
多用途地籍	multi-purpose cadastre

多用途建筑物	multiuse building
多余观测	redundant observation
多余观测数	redundancy number
多圆锥投影	polyconic projection
多源信息融合	multi-source information fusion
多源遥感数据	multi-source remote sensing data
多值的	multi-value
多重分形	multi-fractal

E

额定荷载	nominal load
厄克曼海流计	Ekman current meter
厄特弗效应	Eotvos effect
厄特弗效应改正	Eotvos correction
二叉树	B-tree
二分圈	equinoctial colure
二阶常微分方程	second order ordinary differential equation
二阶导数	second derivative
二阶矩	the second moment
二进制	binary system
二类设计:观测值权的分配问题	second-order design (SOD): the weight problem
二类水体	coastal water
二维空间	two-dimensional space
二维控制场	quasi-2D control field
二向反射测量	bidirectional reflectance measurement
二向反射模型	bidirectional reflectance model
二向性反射	bidirectional reflection
二向性反射分布函数	bidirectional reflectance distribution functions(BRDF)
二向性反射特性	bidirectional reflectance characteristic

二项分布	binomial distribution
二项级数	binomial series
二项式	binomial
二项式定理	binomial theorem
二元光学元件	diffractive optics
二元性	duality
二元一次方程	linear equation in two unknowns
二值图像	binary image
二重根	double root

F

发电	power generation
发光二极管	light-emitting diode(LED)
发光体,光源	illuminant
发散(性)	divergence
发散的	divergent
发散级数	divergent series
发散性迭代	divergent iteration
发射率	emissivity
法方程	normal equation
法国 Spotimage 公司的制图卫星系统	Spot 5
法国发射的用于地球资源遥感的卫星	SPOT (System Probatoite d'observation de La Terre)
法国天泰雷兹	Thales
法截面,法截线	normal section
法截线方位角	normal section azimuth
法截线曲率半径	radius of curvature of normal section
法向量	normal vector
法伊改正	Faye correction
反比例	inversely proportional
反差	contrast
反差系数	contrast coefficient
反差增强	contrast enhancement

反电子欺骗 anti-spoofing(AS)
反对称 anti-symmetric
反对称小波 anti-symmetrical wavelet
反方位角 back azimuth
反光立体镜 reflecting stereoscope
反函数 inverse function
反立体效应 pseudostereoscopy
反三角函数 inverse circular function
反射(性,率,系数) reflectivity
反射,反射比 reflectance
反射光谱 reflectance spectrum
反射率模型 reflectance model
反射率转换 reflectance retrieval
反射器 reflector
反射强度像 intensity image
反束光导管摄像机 return beam vidicon camera(RBV)
反投影变换 back project
反向传播法,逆推学习算法,BP算法 back propagation(BP)
反像 mirror reverse,mirror image
反演方法 retrieval method
反余弦函数 inverse cosine function
反照率 albedo
反正切函数 inverse-tangent function
反正弦函数 inverse-sine function
反证法 proof by contrapositive
反转片 reversal film
反作用(力) reaction (force)

范数,标准,规范　norm
方(矩)阵　square matrix
方案,计划　scheme
方案设计　schematic design
方案设计阶段　schematic design phase
方案研究　plan study
方差　variance
方差分析　analysis of variance
方差膨胀模型　variance inflation model
方差-协方差传播律　variance-covariance propagation law
方差-协方差矩阵　variance-covariance matrix
方差因子　variance factor
方程　equation
方程解　solution of equation
方格纸　graph-paper
方括号　square bracket
方里网　kilometer grid
方位　bearing
方位角　azimuth
方位角闭合差　closure error of azimuth, misclosure in azimuth
方位角测定　azimuth determination
方位角中误差　mean square error of azimuth
方位罗盘　azimuth compass, bearing compass
方位圈,地平经圈,方位度盘　azimuth circle
方位天文学　positional astronomy

方位投影	azimuthal projection
方位向基线	azimuth base-line
方位校准	bearing calibration
方向	direction
方向比	direction ratio
方向改正	correction for direction
方向观测法	method by series, method of direction observation, method of direction observation
方向计算	direction calculation
方向检测	direction checking
方向角	direction angle
方向精度	bearing accuracy
方向可调滤波器	steerable filter
方向亮温	directional brightness temperature
方向平差	adjustment by directions
方向余弦	direction cosine
防洪工程	flood protection works
防火墙	fire barrier
防灾	disaster prevention
房地产	real estate, realty
房地产,权属	property
房地产业	realty industry
仿射变换	affine transformation
仿射纠正	affine rectification
放大	zoom out
放大器	amplifier
放样表	layout table

放样测量	setting-out survey
放样技术	layout technique
飞行高度,航高	flight altitude
飞行器定位	aeroplane location
飞行质量	flying quality
非参数模型	non-parametric model
非常规仪器	non-conventional equipment
非大地测量技术	non-geodetic techniques
非地形摄影测量	nontopographic photogrammetry
非对称	asymmetrical
非对角线元素	off diagonal element
非负的	non-negative
非负矩阵	non-negative matrix
非监督分类	unsupervised classification
非减函数	nondecreasing function
非接触法	non-contact method
非量测相机	non-metric camera
非零特征值	non-zero eigenvalue
非零向量	non-zero vector
非零元素	non-zero element
非平稳过程	nonstationary process
非平稳随机模型	nonstationary stochastic model
非平稳序列	nonstationary series
非奇异方阵	nonsingular square matrix
非奇异阵	nonsingular matrix
非球形粒子	non-spherical particle
非确定性函数	non-deterministic function
非同温像元	non-isothermal pixel

非线性	nonlinear
非线性扩散	nonlinear diffusion
非线性摄动	nonlinear perturbation
非线性数学模型	non-linear math model
非线性退化技术	nonlinear cooling technique
非中心参数	noncentrality parameter
菲列罗公式	Ferreros formula
斐波那契数,黄金分割数	Fibonacci number
斐波那契序列	Fibonacci sequence
费尔马最后定理	Fermat's last theorem
费力	labour intensive
费时	time-consuming
分版原图	flaps
分半法	bisection method
分瓣投影	interrupted projection
分辨率,分辨力	resolution
分辨率增强	resolution enhancement
分布	distribution
分布函数	distribution function
分层	hierarchical
分层处理	hierarchical processing
分层结构影像	hierarchical structure image
分层设色	hypsometric tinting
分层设色表	tint graduation
分层设色法	hypsometric layer
分层提取	layered extraction
分潮	partial tide

分潮迟角	epoch of partial tide
分潮振幅	amplitude of partial tide
分带纠正	zonal rectification
分带子午线	zone dividing meridian
分点岁差	precession of equinox
分段设计	step-by step design
分割,划分	partition
分光辐射计	spectroradiometer
分光技术	spectrum dividing techniques
分洪区	flood diversion area
分画尺	divided scale, graduated scale
分级统计图法	choropleth technique
分角线	angle bisector
分解(遥感尺度转换)	scaling down
分解技术	decomposition technique
分解与重建	decomposition and reconstruction
分块矩阵	partitioned matrix
分类	classification
分类规则	classfication rule
分类精度	classification accuracy
分类决策	classification decision
分类码	classification code
分类器	classifier
分类融合	separate classes fusion
分类压缩	classified compression
分类增强	classified enhancement
分离-合并	separating combining
分力	component of force

F

分量　component
分米　decimeter
分母　denominator
分配律　distributive law
分区规划　district planning
分区密度地图　dasymetric map
分区统计图表法　chorisogram method, cartodiagram method
分区统计图法,等值区域法　cartogram method, choroplethic method
分色,分色参考图　color separation
分数　fraction
分数比例尺　representative fraction(RF)
分数维,分形　fractal
分数维分朗运动场　fractional brownian motion field
分水岭变换　watershed transform
分水区,汇水区域　watershed area
分维　fractal dimension
分位表　distribution table
分位值　distribution value
分析地图　analytical map
分形　fractal
分形布朗运动　fractional Brownian motion
分形测量　fractal measurement
分形几何　fractal geometry
分形压缩　fractal compression
分子　numerator
分组平差　phase adjustment

F

风暴潮　storm surge
风场反演　wind retrieval
风海流　wind-driven current
风速计　anemometer
风速记录仪　anemograph
风险管理　risk management
峰度　kurtosis
峰值方法　peak value method
缝光源　slit light
缝隙　slit
浮标　buoy
浮标式自计验潮仪　buoy autorecording tide gauge
浮点,浮动测标　floating dot
浮雕影像地图　picto-line map
浮动灯标　floating light
浮子验潮仪　float gauge
符号化　symbolization
符合水准器　coincidence bubble
幅,振幅,范围　amplitude
辐射补偿　radiation compensation
辐射潮　radiational tide
辐射传输　radiative transfer
辐射传输方程　radiative transfer equation
辐射传输模型　radiation transfer model
辐射定标　radiative calibration
辐射度量　radiation scaling
辐射改正　radiometric correction
辐射计,光测测定仪,自记曝光计　actinograph

辐射计,射线探测仪	radiometer
辐射亮度	radiance
辐射校正	radiation correction
辐射信噪比	radiation ratio of signal to noise
辐射遥感器	radiation sensor
辐射源识别	emitter recognition
俯角	depression angle
辅助测量杆	aid measuring pole
辅助导线	auxillary traverse
辅助方程	auxiliary equation
辅助观测船	auxiliary observing vessel
辅助角	auxiliary angle, subsidiary angle
辅助设备	auxiliary equipment
辅助数据	auxiliary data
辅助圆	auxiliary circle
负,负片	negative
负荷	load
负角	negative angle
负矢量	negative vector
负数	negative number
负整数	negative integer
附参数条件平差	condition adjustment with parameters
附合导线	connecting traverse
附合水准路线	annexed leveling line
附加参数	additional parameter's(APs)
附加费用	additional charge
附加控制点	additional control point

F

附条件参数平差，附条件间接平差　parameter constrained adjustment
附属构筑物　ancillary structure
附属建筑物　ancillary buliding
附属设施　ancillary facility
复测法　repetition method
复共轭　complex conjugate
复合分类　mixed classification method
复合函数　composite function
复合数　composite number
复角　compound angle
复垦测量　reclaimation survey
复曲线　compound curve
复数　complex number
复数的辐角　argument of a complex number
复数根　complex root
复数平面　complex number plane
复数小波变换　complex wavelet transform
复相关函数　complex correlation function
复杂多边形　complex polygon
复杂目标　complex object
复杂要素　complex feature
副台　slave station
傅立叶-梅林变换　Fourier-Mellin transform
傅立叶变换　Fourier transform
傅立叶逆变换　inverse Fourier transformation

G

伽利略控制中心	GALILEO control centers (GCC)
伽利略卫星定位系统	Galileo positioning system
伽利略系统[欧洲]	GALILEO
伽瓦德	Javad
改化后的重力值	reduced gravity
改化经度	reduced longitude
改化距离	reduced distance
改建	alteration
改建地区	improvement area
概率	probability
概率乘法定律	multiplication law (of probability)
概率分布	probability distribution
概率距离	probability distance
概率空间	probability space
概率论	law of probability, theory of probability
概率密度函数	probability density function (PDF)
概率母函数	probability generating function
概率判决函数	probability decision function
概率水平	probability level
概率松弛法	probability relaxation algorithm
概略地貌	approximate relief
概略定向	approximate orientation
概念模型	conceptual data model

概然误差,或然误差	probable error
概算三角形	priliminary triangle
干出滩	drying rock
干旱	drought
干旱区	arid areas
干旱预警模型	drought prediction model
干涉侧扫声呐	interference sidescan sonar
干涉合成孔径雷达,雷达干涉测量	interferometric synthetic aperture radar(InSAR)
干涉技术	interferometry
干涉雷达	interferometry SAR
干涉条纹	interferometric fringe
干涉条纹图	fringe
干涉图	interferogram
干湿球温度表	psychrometer
感光	sensitising, sensitization
感光测定	sensitometry
感光特性曲线	characteristic curve of photographic emulsion
感知规律	perception law
刚结构	rigid frame
刚体	rigid body
钢尺	steel tape
钢筋混凝土	reinforced concret(RC)
港口	port
港口城市	harbor city
港口工程测量	harbor engineering survey
港湾测量	harbor survey

港湾海岸	embayed coast
港湾图	harbor chart
高层建筑	high-rise building
高层建筑物振动测量	measurement of vibrations of tall buildings
高差	difference of height, elevation difference
高潮	high water
高潮高	height of high water
高潮时	time of high water
高程	elevation, height
高程测量，垂直测量	vertical survey
高程差	difference of elevation
高程传递	transfer of elevation
高程点	altimetric point, elevation point
高程改正	altitude correction
高程基准	height datum, vertical datum
高程计算	elevation calculation
高程精度	elevation accuracy, height accuracy, vertical accuracy
高程控制	elevation control
高程控制测量	vertical control survey
高程控制点	vertical control point, vertical control station
高程控制加密	vertical control densification
高程控制精度	accuracy of vertical control
高程控制网	vertical control network
高程数据	altimetric data

G

高程系统	height system
高程异常	height anomaly
高程原点	height datum
高程中误差	mean square error of height
高程注记	elevation notation
高等三角测量	higher-order triangulation
高度,地平纬度	altitude
高度角	altitude angle, elevation angle
高度平均误差	average error in altitude
高度视差	parallax in altitude
高对比度	high-contrast
高分辨率	high resolution
高分辨率航空影像	high resolution aerial image
高分辨率可见光传感器(SPOT 卫星搭载)	HRV(High Resolution Visible)
高分辨率全色影像	high resolution spatial panchromatic image
高分辨率摄像机	high resolution cameras
高光谱	hyperspectral imagery
高光谱反射率	hyperspectral reflectance
高光谱图像	high spectral image
高光谱纹理	hyperspectral texture
高光谱遥感	hyperspectral remote sensing
高光谱影像	hyperspectral image
高精度定位算子	high-precision location operator
高精度轮廓	high-accurate outline
高精度水准测量	high-precision leveling
高空间分辨率	high spatial resolution

G

高密度数字磁带	high density digital tape(HDDT)
高斯分布,正态分布	Gaussian distribution
高斯-克吕格坐标	Gauss-Krüger coordinates
高斯-马尔柯夫模型	Gauss-Markoff model
高斯模糊	gauss blur
高斯平面坐标	Gauss plane coordinate
高斯平面坐标系	Gauss plane coordinate system
高斯投影	Gauss projection
高斯投影方向改正	arc-to-chord correction in Gauss projection
高斯投影距离改正	distance correction in Gauss projection
高斯-克吕格投影	Gauss-Krüger projection
高速公路	express way,high speed highway
高通滤波	high frequency modulation, high-pass filtering
高通滤波器	high-pass filter
高纬度	high latitude
高压电	high-voltage electricity
高原	upland plain
割点	point of secant
格林尼治平恒星时	Greenwich mean sidereal time (GMST)
格林尼治真恒星时	Greenwich apparent sidereal time (GAST)
格林尼治子午线,起始子午线	Greenwich meridian
格林尼治时间	Greenwich mean time

G

格式转换　format conversion
格网单元　cell
格网单元尺寸　cell size
格网简化　mesh simplification
各向异性扩散　anisotropic diffusion
给水工程,供水工程　water supply engineering
跟踪　tracing
跟踪精度　tracking accuracy
跟踪台　tracking station
跟踪仪　tracker
更新　update
更新速率　update rate
工厂区　factory area
工厂现状图测量　survey of present state at industrial site
工程测量　engineering survey
工程测量学　engineering surveying(geodesy)
工程地质勘察　engineering geological investigation
工程地质学　engineering geology
工程概况　project profile
工程建筑物　engineering structures
工程经纬仪　engineer's theodolite
工程决算　final building cost
工程科学　engineering sciences
工程控制网　engineering control network
工程设计　engineering design
工程摄影测量　engineering photogrammetry
工程水准仪　engineer's level

工程项目	engineering project, engineering drafting
工程制图	engineering drawings
工地(总)平面图	site plan, site map
工地,现场	site
工业测量系统	industrial measuring system
工业城市	industrial city
工业建筑	industrial architecture
工业区	industrial district
工业摄影测量	industrial photogrammetry
工业突发事故	industrial sudden accidents
工业污水	industrial sewage
工业用地	industrial land
工作流程	workflow
工作站	workstation
公差	common difference
公共建筑	public building
公共绿地	green public space
公海	international waters
公斤,千克	kilogramme
公里,千米	kilometer
公理	axiom
公路,道路	highway
公路定线	route alignment
公路桥	highway bridge
公路设计	highway design
公路网	highway network
公切	common tangent

公式	formula(formulae)
公弦	common chord
功率	power
功率谱	power spectrum
拱门，弓形结构，拱形	arch
拱桥	arch bridge
拱座,弹簧座	spring block
共轭	conjugation
共轭(直)径	conjugate diameters
共轭复数	complex conjugate
共轭双曲	conjugate hyperbola
共轭轴	conjugate axis
共面	coplanar
共面方程	coplanarity equation
共面力	coplanar forces
共面条件	coplanar conditions
共面	coplanar lines
共生矩阵	co-occurrence matrix
共线	collinearity
共线方程	collinearity equation
共线面	collinear planes
共圆	concentric
共轴	coaxial
共轴系	coaxial system
共轴圆	coaxial circles
构像方程	imaging equation
构造变形	tectonic deformation

构造现象	tectonic phenomena
构造运动	tectonic motion
估计,估计量	estimate
估计,估算	estimation
估计到达时间(导航)	estimated time of arrival (ETA)
估计位置误差	estimated position error (EPE)
估值	estimate value
孤立建筑物	detached building
孤立节点	isolated node
古地图	ancient map
古建筑测绘	old architecture survey
古建筑与古文物摄影测量	architectural and archaeological Photogrammetry
骨架航线	control strip
骨架化	skeleton
骨架线	skeleton line
鼓扫描器	drum scanner
固定点	fixed point
固定平极	fixed mean pole
固定起始值,固定基准	fixed datum
固定天线	fixed antenna
固定误差	fixed error
固定相移	fixed phase drift
固态摄像机	solid state camera
固体潮	solid Earth tide
固体潮观测	Earth tide observation
固体激光器	solid-state laser

固有误差 inherent error
故意破坏行为 vandalism
故障,缺陷,干扰,雷达位置测定器,窃听器 bug
顾问委员会 advisory committee
挂图 mural map
拐点 inflection point
关键帧 key frames
关键字 key word
关联规则 spatiotemporal association rule
关系匹配 relational matching
观测方案 observation plan
观测方程 observation equation
观测基线 observation base
观测精度 measurement accuracy, observation accuracy
观测天文学 observational astronomy
观测误差 observation error
观测值 observation, observed value
观测值的随机特性 stochastic properties of the observations
观测值的正态性 normality of observations
观测值相关性 correlation of observations
观测值协方差矩阵 covariance matrix of the observables
观测重力值 observed gravity
观象台、天文台 observatory

冠层反射模型 canopy reflectance model
管道,管线 pipeline
管道测量 pipe survey
管道定线 pipe alignment
管道系统 piping system
管网 pipeline network
管线测量 pipeline survey
贯通测量 breakthrough survey
贯通点 breakthrough point
贯通精度 accuracy of break through, breakthrough accuracy
惯性矩 moment of inertia
惯性 inertia
惯性参考设备 inertial reference unit(IRU)
惯性参考系统 inertial reference system
惯性测量系统 inertial surveying system(ISS)
惯性测量装置 IMU
惯性导航系统 inertial navigation system(INS)
惯性空间 inertial space
惯性坐标系 inertial coordinate system
惯用名 conventional name
光斑,光点 illuminated patch
光点地震仪 optical point seismometer
光电测距仪 electro-optical distance measuring instrument
光电管、光电池、光电元件 photocell
光电遥感器 photoelectric sensor

光度计	aerolux
光流	optical flow
光谱	spectrum
光谱保持	spectral preservation
光谱变量	spectral variable
光谱测量技术	spectral measurement methods
光谱的	spectral
光谱范围	spectral range
光谱分解	spectral decomposition
光谱分析	spectral analysis
光谱感光度,光谱灵敏度	spectral sensitivity
光谱混合	spectral mixing
光谱距离	spectral distance
光谱模拟	spectral simulation
光谱匹配	spectral matching
光谱匹配技术	spectral matching technique
光谱曲线	spectral profile
光谱特性	spectral characteristic
光谱特征	spectral signature
光谱吸收	spectral absorption
光谱响应	spectral response
光谱信息特征	spectral information characteristics
光谱形状	spectral shape
光谱指数	spectral indices
光谱重建	spectral re-building
光圈号数	f-number, stop-number
光束,射束	bundle of rays

光束法空中三角测量　bundle aerial triangulation
光束法平差　bundle adjustment
光速测距仪,光电测距仪　geodimeter
光纤　optical fiber
光学补偿器　optical compensator
光学测距仪　optical rangefinder
光学传递函数　optical transfer function(OTF)
光学传感器　optical sensor
光学对中器　optical plummet
光学经纬仪　optical theodolite
光学密度　optical density
光学扫描仪　optical scanner
光学水准仪　optical level
光学条件　optical condition
光学投影　optical projection
光学系统　optical train
光学相关　optical correlation
光学影像　optical image
光学准直　optical alignment
光栅　grating
广播星历　broadcast ephemeris
广义点摄影测量　generalized point photogrammetry
广义积分　improper integral
广义空间信息网格　generalized spatialized information grid
广义岭估计　generalized ridge estimate
广义逆　generalized inverse

广义判别分析 generalized discriminate analysis
广义平差 generalized adjustment
广义相对论 general relativity
广域差分 GPS wide area differential GPS (WADGPS)
广域增强系统 wide area augmentation system (WAAS)
归化纬度 reduced latitude
归零,零位调整,零点调整 zero adjustment
归算 reduction
归心改正 correction for centering
归心计算 eccentric reduction
归心元素 elements of centring
归一化极化指数 normalized polarization index (NDPI)
归一化荧光高度 normalized fluorescence height
归一化植被指数 normalized difference vegetation index(NDVI)
归约公式 reduction formula
规定航向 prescribed course
规定限差 accepted tolerance
规范,规程,施工说明书 specification
规划标准 planning criteria
规划程序 planning procedure
规划调查 planning survey
规划审批程序 procedure for approval of urban plan

规划图	planning map
规划许可	planning permission
规则格网模型	regular square grid
规则格网数字高程模型	regular grid DEMs
规则化	regularization
轨道	orbit
轨道参数	orbit parameters
轨道观测卫星系统，美国轨道科学公司（Orbimage）2001 年发射	Orbview 4
轨道岁差	orbital precession
轨迹	locus
轨迹方程	equation of locus
滚轮算法	roller wheel algorithm
国际 GNSS 服务组织	international GNSS service（IGS）
国际 GPS 数据处理结果的交换格式	software independent exchange (format)(SINEX)
国际标准化组织	International Organization for Standardization(ISO)
国际标准化组织地理信息/地球信息技术委员会	Geographic Information/Geomatics,ISO,ISO/TC 211
国际测量师联合会	International Federation of Surveyors(FIG)
国际大地测量协会	International Association of Geodesy(IAG)

国际大地测量与地球物理联合会	International Union of Geodesy and Geophysics(IUGG)
国际地理学联合会	International Geographical Union (IGU)
国际地球参考框架	international terrestrial reference frame(ITRF)
国际地球参考系	international terrestrial reference system (ITRS)
国际地球自转服务	International Earth Rotation and Reference Service (IERS)
国际海床管理局	International Sea Bed Authority
国际海道测量局	International Hydrographic Bureau
国际海道测量组织	International Hydrographic Organization(IHO)
国际海道测量组织东大洋海洋测量委员会	East Atlantic Hydrographic Commission
国际海道测量组织东亚海道测量委员会	East Asia Hydrographic Commission
国际海道测量组织海图标准化委员会	Chart Standardization Committee of IHO
国际海道测量组织数字数据交换委员会	Committee on the Exchange of Digital Data
国际海道测量组织制定的电子海图显示与信息系统性能标准草案	draft performance standard
国际海底地形图	international bathymetric chart
国际海上无线电通讯委员会	International Radio Maritime committee

国际海事组织	International Maritime Organization (IMO)
国际海图	international chart
国际海洋考察理事会	International Council for Exploration of the Sea
国际海洋物理科学协会	International Association for the Physical Science of the Ocean
国际科学(联合会)理事会	International Council for Science, International Council of Scientific Unions(ICSU)
国际摄影测量与遥感学会	International Society for Photogrammetry and Remote Sensing (ISPRS)
国际天球参考架	international celestial reference frame(ICRF)
国际天文学联合会	International Astronomical Union (IAU)
国际椭球	international spheroid
国际协议原点	conventional international origin (CIO)
国际原子时	international atomic time(TAI)
国际制图协会	International Cartographic Association(ICA)
国家 GPS 大地控制网	national GPS control network
国家参考框架	national reference frame
国家测绘局	State Bureau of Surveying and Mapping (SBSM)
国家地图集	national atlas

G

国家高程基准	national vertical datum
国家高程控制网	national vertical control network
国家基础地理信息系统	national fundamental geographic information system
国家空间数据基础设施	national spatial data infrastructure (NSDI)
国家平面控制网	national horizontal control network
国家水准网	national leveling network
国家重力基本网	national basic gravity network
国民生产总值	gross national product(GNP)
国内生产总值	gross domestic product(GDP)
国土规划	national land planning, territorial planning
过渡曲线	transition curves
过街人行道	crosswalk

H

海岸	sea coast
海岸带	coastal zone
海岸带地形测量	littoral topographic survey
海岸带地形图	littoral zone topographic chart
海岸地貌	coastal landform
海岸地形测量	coastal topographic survey
海岸图	coastal chart
海岸线	coast line
海岸性质	nature of the coast
海拔高程	sea-level elevation
海滨	seashore
海槽	trough
海道测量,水道测量	hydrographic survey
海道测量委员会	Hydrographic Commission
海底成像系统	seafloor imaging system
海底底质图	chart of bottom quality
海底地貌探测仪	bottom sounder
海底地貌图	submarine geographic chart
海底地势	submarine relief
海底地势图	submarine situation chart
海底地形	seafloor topography
海底地形测量	bathymetric surveying
海底地形图	bathymetric chart
海底地质构造图	submarine structure chart

H

海底电缆　submarine cable
海底高原　oceanic plateau
海底管道　submarine pipeline
海底火山　submarine volcano
海底控制网　submarine control network
海底面状探测　seabed feature survey
海底倾斜改正　seafloor slope correction
海底山脉　submarine range
海底声标　seafloor acoustic beacon
海底施工测量　submarine construction survey
海底隧道测量　submarine tunnel survey
海底峡谷　submarine canyon
海底重力仪　ocean bottom gravimeter, seafloor gravimeter
海沟　trench
海槛　sill
海军导航卫星系统　navy navigation satellite system (NNSS)
海军海道测量局　Navy Hydrographic Office
海军海洋局　Oceanographic Office
海控点　hydrographic control point
海拉瓦的混合数字摄影测量系统　DCCS(Digital Comparator Correlation System)
海浪　ocean wave
海浪图　sea swell chart
海浪学　science of ocean wave
海浪预报　wave forecasting
海浪预报图　wave forecasting chart

海里(1 海里 =1 852 米)	nautical Mile
海量数据	massive data set
海岭	submarine ridge
海流	sea current
海流观测	current observation
海流计	current meter
海流图	current chart
海流学	current theory
海隆	rise
海幔	apron
海面地形	ocean topography, sea surface topography(SST)
海面地形大气改正	atmospheric correction of sea surface topography
海面地形模型	sea surface topography model
海面高	sea surface height
海面观测	sea survey
海面回波	sea echo
海盆	oceanic basin
海平面基准面	sea-level datum
海丘	knoll
海区界线	sea area bounding line
海区图	regional chart
海区总图	general chart of the sea
海上导航	marine navigation
海上平台	offshore platform
海上油井	offshore oil well
海水密度	density of sea water

H

海水密度图　water density chart
海水声速　sound velocity in water
海水透明度图　water transparency chart
海水温度　temperature of sea water
海水温度图　water temperature chart
海水现场密度　density of local sea water
海水盐度　salinity of sea water
海水盐度图　water salinity chart
海水状态方程　equation of state of sea water
海图　chart
海图比例尺　chart scale
海图编号　chart numbering
海图编绘　composite drawing chart
海图编制　chart compilation
海图标记　chart lettering
海图标题　chart title
海图大改正　chart large correction
海图分幅　chart subdivision
海图符号　chart symbols
海图改正　chart correction
海图更新　chart revision, chart update
海图基准面　chart datum
海图集　chart atlas, marine atlas
海图精度　chart accuracy
海图刻图　chart scribing
海图内容　chart content
海图数据库　chart data base
海图投影　chart projection

海图图廓	chart boarder
海图图式	symbols and abbreviations on chart
海图小改正	chart small correction
海图制图	charting
海图制印	chart reproduction
海图种类	classification of charts
海图自动改正	automatic chart correction
海图自动制图系统	automated cartography system for charting
海湾	bay, gulf, bight
海峡	strait
海洋测绘	hydrographic surveying and charting, marine charting
海洋测绘数据库	marine charting database
海洋测量	marine survey
海洋测量标志	hydrographic mark
海洋测量定位	marine survey positioning
海洋磁力测量	marine magnetic survey
海洋磁力仪扫海	marine magnetometer sweeping
海洋磁力异常	marine magnetic anomaly
海洋磁力异常图	marine magnetic anomaly chart
海洋大地测量	marine geodesy
海洋大地测量控制网	marine geodetic control network
海洋岛	ocean island
海洋地磁图	marine magnetic chart
海洋地理学	marine geography
海洋地球物理图	marine geophysical chart
海洋地震图	marine earthquake chart

H

海洋地质调查	marine geological survey
海洋地质学	marine geology
海洋电磁学	marine electromagnetics
海洋调查	oceanographic survey
海洋调查规范	the specification for oceanographic survey
海洋浮标	ocean buoy
海洋工程测量	marine engineering survey
海洋观测	oceanic observation
海洋光学仪器	marine optical instrument
海洋航空遥感	marine airborne remote sensing
海洋划界测量	marine demarcation survey
海洋环境图	marine environmental chart
海洋气候	marine climate
海洋气象图	marine meteorological chart
海洋气象学	marine meteorology
海洋全磁力图	total magnetic intensity chart
海洋铯光泵磁力仪	marine cesium magnetometer
海洋生态学	marine ecology
海洋生物图	marine biological chart
海洋声学	marine acoustics
海洋声学技术	marine acoustic techniques
海洋水文图	marine hydrological chart, ocean hydrological chart
海洋水准测量	marine leveling
海洋卫星	seasat
海洋学	oceanography
海洋研究科学委员会	Scientific Committee on Oceanic Research(SCOR)

海洋研究特别委员会	special committee for oceanographic research
海洋遥感	ocean remote sensing
海洋灾害	ocean disaster
海洋噪声	sea noise
海洋质子采样器	marine bottom proton sampler
海洋质子磁力仪	marine proton magnetometer
海洋重力	marine gravity
海洋重力测量	marine gravimetry
海洋重力测量基点	marine gravity base point
海洋重力测线布设	marine gravimetry traverse line design
海洋重力仪	marine gravimeter
海洋重力异常	marine gravity anomaly
海洋重力异常图	marine gravity anomaly chart
海洋专题测量	marine thematic survey
海洋资源图	marine resource chart
海运安全组织	Sub-Committee on Safety of Navigation
函数模型	functional model
旱地	dry land
航标	aid to navigation
航测地形图	aerial topographic map
航测平面图	aerophotographic plan
航测设备	aerial equipment
航测摄影	aerial mapping photography
航测摄影机	aerial surveying camera
航程	distance run

H

航带法空中三角测量	strip aerial triangulation
航带拼接	strip mosaic
航道	fairway
航道,通道	channel
航道图	navigation channel chart
航道校正,轨道调整	track adjustment
航高	flight altitude
航海天文历	nautical almanac
航海通告	notice to mariners
航海图	nautical chart
航迹绘算	course plotting
航迹推算	track estimation
航迹自绘仪	track plotter
航空测图仪	aerocartograph
航空导航图	aeronautical chart
航空的,航空导航的	aeronautical
航空的,空中的,天线	aerial
航空平台	aerial platform
航空摄谱仪	aerial spectrograph
航空摄影	aerial photography
航空摄影测量	aerial photogrammetry, aerophotogrammetric survey, aerophotogrammetry
航空摄影测量,航空摄影大地测量	aerophotogeodesy
航空摄影测量的,航测的	aerophotogrammetric(al)
航空摄影测量学	aerial photogrammetry

H

航空摄影测深系统	aerophotogrammetry bathymetric system
航空摄影的,航摄的	aerophotographic(al)
航空摄影机	aerial camera, aerophotographic camera
航空摄影学,航空摄影	aerophotography
航空摄影制图	aerial photomapping
航空摄影装置	aerial camera mounting
航空数码相机	aerial digital camera
航空水准测量,航空抄平	aerial leveling
航空像片,航摄像片	aerophotograph
航空遥感	aerial remote sensing, airborne remote sensing
航空影像	aerial image
航空制图	aeronautical charting
航空重力测量	airborne gravimetry
航路点	waypoint
航路指南	sailing directions
航片	aerial photo
航偏角	angle of drift
航摄负片	aerophotographic negative
航摄规范	aerial photography norm
航摄计划	flight plan of aerial photography
航摄胶片	aero-photograph film
航摄勘测	aerophotographic reconnaissance
航摄领航	navigation of aerial photography

H

航摄漏洞	aerial photographic gap
航摄面	aerial surface
航摄软片	aerial film
航摄图	aerial map
航摄图片判读	aerial photo interpretation
航摄仪,航空摄影机	aerial photographic camera, aerial surveying camera, aircraft camera
航摄仪镜头	aerial camera lens
航摄仪校准焦距	aerial-camera-calibrated focal length
航摄质量	quality of aerophotography
航摄资料地区	aerial coverage
航天飞机	space shuttle
航天摄影	space photography
航天摄影测量,太空摄影测量	space photogrammetry
航天遥感	space remote sensing
航天站	space station
航线,线路	route
航线测量	aerial strip survey
航线法区域网平差	block adjustment by strips
航线反转	inverse route
航线偏航指示	course deviation indicator (CDI)
航线图	aerial route map
航向	heading
航向,航迹	track (TRK)
航向倾角	longitudinal tilt, pitch
航向向上显示	track-up display

H

航向重叠	fore-and-aft overlap
航行三角形	course triangle
航行图	sailing chart
航行障碍物	navigation obstruction
航用星	navigation star
毫克	milligramme(mg)
毫米	millimeter(mm)
毫米波	mm-wave
毫升	milliliter(ml)
合成地图	synthetic map
合成概率	compound probability
合成孔径雷达	synthetic aperture radar(SAR)
合成孔径声呐	synthetic aperture sonar
合点,灭点	vanishing point
合点控制	vanishing point control
合速度	resultant velocity
河床	riverbed
河道整治测量	river improment survey
核磁共振计算机断层扫描仪	nuclear magnetic resonance (NMR)
核点	epipole
核面	epipolar plane
核模	Slichter modes(triplet)
核曲线	epipolar curve
核算	check computation
核线	epipolar line,epipolar ray
核线相关	epipolar correlation
核轴	epipolar axis

盒式分类法	box classification method
盒维数	box dimension
赫尔默特椭球	Helmert'spheroid
赫耳德不等式	Holder's inequality
黑板框架	blackboard architecture
黑体	black-body
恒力	constant force
恒速	constant velocity
恒速率	constant speed
恒星跟踪器	STA(star tracker)
恒星检校	steller calibration
恒星日	sidereal day
恒星摄影机	stellar camera
恒星时	sidereal time
恒星视差	stellar parallax
桁架,桁梁,构架	truss
桁架跨度	truss span
桁架梁	trussed beam
横断面,断面图,剖面图	cross section
横断面测量	cross-section survey
横断面水准测量	cross-section leveling
横断面图	cross-section drawing
横截分量	transverse component
横坑,入口	adit
横倾	list
横向比例畸变	transverse proportion distortion
横向分辨率	transverse resolution

H

横向海岸	transverse coast
横轴	transverse axis
横轴墨卡托投影地图	transverse Mercator chart
横轴投影	transverse projection
横坐标	abscissa
横坐标轴,X 轴	abscissa axis
红边参数	red edge parameter
红外(热)辐射	IR (heat) radiation
红外波谱	infrared spectrum
红外测距仪	infrared EDM instrument, infrared ranger
红外窗口	infrared window
红外辐射	infrared radiation
红外辐射计	infrared radiometer
红外片	infrared film
红外扫描仪	infrared scanner
红外摄影	infrared photography
红外图像	infrared imagery
红外卫星云图	infrared cloud imagery
红外遥感	infrared remote sensing
洪涝灾害	flood disaster
洪水监测	flood monitoring
洪水控制	flood control
后尺	back rod
后处理 GPS	post-processed GPS
后处理差分改正	post-processed differential correction
后处理动态	post-processed kinematic(PPK)

H

后方交会	resection
后视	backsight(BS)
后向反射	back reflectance
后向散射模型	backscatter model
后向散射系数	backscattering coefficient
后验	posteriori
后验方差因子	posteriori variance factor
后验概率	posterior probability
弧长	arc length
弧度	radian
弧度测量	arc measurement
弧度法	circular measure
弧段	arc
弧-结点拓扑关系	arc-node topology
湖泊沉积	lake sediment
虎克定律	Hooke's law
互补概率	complementary probability
互补色地图	anaglyphic map
互补色镜	anaglyphoscope
互补色立体测图仪	anaglyphic plotter
互补色立体观察	anaglyphical stereoscopic viewing
互补事件	complementary event
互补相关	complementary relationship
互不相交	mutually disjoint
互操作	interoperability
互斥事件	exclusive events
互相关	cross correlations
互协方差	cross covariance

互易换能器	reciprocal transducer
护岸,护堤	bank paving
滑动平均参数	moving average parameter
滑动平均模型	moving average model
滑动式投影仪	slide projector
滑轮	pulley
滑模施工	slip forming
滑坡	landslide
滑坡监测	landslide monitoring
划定界限	boundary settlement
环境地图	environmental map
环境分析	environmental analysis
环境管理	environmental management
环境规划	environmental planning
环境监测	environmental monitoring
环境建筑学	environmental architecture
环境评估	environmental asessment(EA)
环境探测卫星	environmental survey satellite
环境污染	environmental pollution
环境因子	environmental factors
环线闭合差	circuit closure
环形公路,环路	circuit highway
环形铁路	circuit railroad
缓冲区	buffer
缓冲区分析	buffer analysis
缓和曲线	easement curve
缓和曲线测设	spiral curve location
换能器	transducer

换能器吃水改正	correction of transducer draft
换能器动态吃水	transducer dynamic draft
换能器基线改正	correction of transducer baseline
换能器静态吃水	transducer static draft
换能器拖体	transducer tow vehicle
换能器校准	calibration of transducer
换能器阵列	base array
换算,转换	conversion
换算因子	conversion factor
荒漠化	desertification
黄道	ecliptic
黄海平均海[水]面	Huanghai mean sea level
黄极	ecliptic pole
黄金分割	golden section
黄经	ecliptic longitude
黄纬	ecliptic latitude
灰度	grey level
灰度不均匀校正	uneven grey rectification
灰度共生矩阵	grey level co-occurrence matrix (GLCM)
灰度匹配	area based image matching
灰度剖面分析	analysis of image profile
灰度形态学	grey level morphology
灰度值	gray(= grey) value
灰关联度	grey correlation grade
灰关联分析	grey correlation analyses
灰集理论	grey sets
灰色模型	grey modeling

H

灰色色调,黑白亮度等级	Gray tone
灰色系统	grey system
灰色系统理论	grey system theory
灰楔	grey wedge, optical wedge
恢复,修复	restoration
回波	return wave
回波测高仪	echo altimeter
回波信号	echo signal
回代	back substitution
回光反射标志	retro-reflective targets(RRT)
回声测深	echo sounding
回声测深仪	echo sounder
回声测深仪总改正	total correction of echo sounder
回声测声器,回声探测器	acoustic sounder,echo sounder
回声扫测仪	echo sweeper
回声探测	acoustic sounding
回填,再填	refill
回头曲线	reverse curve
回头曲线测设	hair-pin curve location
回旋螺旋曲线	clothoid spiral curve
回转半径	radius of gyration
回转流	rotary current
回转天线,旋转天线	rotating aerial
汇水面积	catchment area
汇水面积测量	catchment area survey
绘图,展绘	plot

H

绘图机	plotter
绘图精度	accuracy of drawing, drafting accuracy
绘图文件	plotting file
绘图桌,量测台	plotting table
混叠	aliasing
混沌免疫算法	chaos immune algorithm
混合采样	composite sampling
混合潮	compound tide
混合潮港	mixed tide harbor
混合反演	hybrid inversion
混合光谱	mixed spectrum
混合结构	hybrid structure
混合像元	mixed image cells, mixed pixel
混合像元分解	decomposing method of mixed pixel
混合形式的数字高程模型	grid-TIN
混凝土	concrete
混凝土标石	concrete pillar, concrete post
混凝土衬砌	concrete lining
活动(地质)构造变形	active tectonic deformation
活动觇牌法	method of moving target
活动轮廓	snake model
火山爆发	volcano eruptions
火山岛	volcanic island
火灾	fire accident
火灾报警系统	fire alarm system

H

J

机场识别	airport recognition
机器人系统	robotic system
机载侧视雷达，侧视雷达	side-looking airborne radar (SLAR)
机载磁测深系统	airborne magnetic bathymetric system
机载多光谱遥感测深系统	airborne multispetral remote sensing
机载合成孔径雷达	airborne SAR
机载激光测深系统	airborne laser sounding system
机载激光测深	airborne laser sounding
机载激光测图	airborne laser mapping
机载激光成像雷达扫描成像模型	scan mode of airborne-based image laser radar
机载激光地形测图	airborne laser terrain mapping (ALTM)
机载激光扫描	airborne laser scanning
机载三线阵 CCD	airborne TLS CCD
机载遥感器	airborne senor
机助测图	computer-assisted mapping, computer-assisted plotting, computeraided mapping
机助地图制图	computer-aided cartography (CAC), computer-assisted cartography (CAC)

机助分类	computer-assisted classification
积集	product set
基	basis
基本[控制]网	fundamental network
基本比例尺	basic scale
基本比例尺,主比例尺,游标比例尺	primary scale
基本大地测量	basic geodetic survey
基本导线,主导线	principal traverse
基本控制网	basic network,fundamental network
基本三角网	basic triangulation network
基本事件	elementary event
基本图形元素	primary graphic elements
基础,地基	foundation(footing)
基础沉降(陷)	foundation settlement
基础底图	base map
基础工程,地基工程	foundation works
基础荷载	foundation load
基础平面图	foundation plan
基础天体测量学	foundamental astrometry
基带	base-band
基点	base point
基-高比	base-height ratio
基坑	foundation pit
基线	baseline
基线测量	baseline measurement
基线长度	baseline length

J

基线方向	baseline direction
基线改正	baseline correction
基线距离比	base/distance ratio
基线平差	baseline adjustment
基线网	baseline network
基线荧光高度	fluorescence line height
基于边缘的匹配	edge-based matching
基于梯度的自适应滤波	gradient-based adaptive filter
基于位置服务	location-based services(LBS)
基准点	datum point
基准面	datum
基准频率	fundamental frequency
基准网	fiducial network
基准误差	datum error
基准线	datum line
基准线	reference line
基准站	base station
基准中央子午线,基准中央经线	reference central meridian
基座	footing
畸变	distortion
畸变差	distortion error
激发极化	induced polarization(IP)
激光测高	laser altimetry
激光测高仪	laser altimeter
激光测距	laser ranging

J

激光测距仪 laser distance measuring instrument, laser ranger, laser rangefinder, laser ranger, laser distance measuring instrument
激光测深 laser bathymetry
激光测深仪 laser sounder
激光测月 lunar laser ranging(LLR)
激光成像雷达 laser image radar
激光点 laser dot
激光定轨 laser orbit determination
激光发射器 laser transmitter
激光绘图机 laser plotter
激光脚点 footprint
激光经纬仪 laser theodolite
激光脉冲频率 laser pulse rate
激光目镜 laser eyepiece
激光铅直仪，激光对中仪 laser plummet
激光扫描 laser scanning
激光扫描测距 scanning laser rangefinder
激光扫描仪 laser scanner
激光束 laser beam
激光水准仪 laser level
激光探测与测距（激光雷达） light detection and ranging (LIDAR)
激光投点 laser plumbing
激光陀螺仪 laser gyroscope

激光照排机	image setter
激光准直	laser alignment
激光准直仪	laser alignment equipment, laser aligner
激活函数	activation function
激活航线	active leg
奇函数	odd function
奇偶嵌入法	odd-even embedding
奇数	odd number
吉奥地理信息管理软件	GeoStar
极	pole
极大	maximium
极地	polar region
极化分析	polarimetric analysis
极化干涉	polarimetric SAR interferometry
极化雷达	polarimetric radar
极球面投影	polar stereographic projection
极三角形	polar triangle
极限	limit
极限定理	limit theorems
极限误差	limit error
极小	minimum
极小值	minimum value
极移	polar motion
极移改正	reduction for polar motion
极值	extreme value, extremum
极值点	extreme point

J

极轴	polar axis,pole axis
极坐标	polar coordinate
极坐标定位,距离方位定位	point coordinate positioning
极坐标量测仪	polar comparator
极坐标缩放仪	polar pantograph
极坐标系统	polar coordinate system
集,测回	set
集成定位系统	integrated positioning systems
集成数据	integrated data
集群	cluster
几何,几何学	geometry
几何参数	geometric parameters
几何大地测量学	geometric geodesy
几何定向	geometric orientation
几何反转原理	principle of geometric reverse
几何分布	geometric distribution
几何光学	geometric optics
几何光学模型	geometric-optics model
几何畸变	geometric distortion
几何建模	geometric modeling
几何精度	geometric accuracy
几何精度因子	geometric dilution of precision (GDOP)
几何精校正	geometric precise correction
几何模型	geometric model
几何配准	geometric co-registration
几何匹配	geometry matching

J

几何水准测量	direct leveling, spirit leveling
几何特征	geometric feature
几何条件	geometric condition
几何校正	geometric correction, geometric rectification
几何形状特征	shape parameter
计程仪,里程表	odometer
计划航向	intended course
计曲线,加粗等高线	index contour
计算机地图制图	computer cartography
计算机断层成像	CT
计算机断层扫描仪	computer tomography scanner
计算机辅助测图/机助测图	computer aided mapping(CAM)
计算机辅助设计	computer aided design(CAD)
计算机兼容磁带	computer compatible tape(CCT)
计算机科学	computer science
计算机视觉	computer vision
计算机体断层成像	computed tomography(CT)
计算机图形学	computer graphics
计算机制图	computer mapping
计算机制图综合	computer cartographic generalization
计算器	calculator
记录格式	record format
技术报告	technical report
技术规范	technical specification
技术术语	technogical term

J

加常数	addition constant
加法	addition
加密	densification
加密[控制]网	densification network
加密测深线	interline of sounding
加密点	pass point
加密码	Y-code, encrypted code
加拿大的实时摄影测量系统	IRI-D256
加拿大的卫星遥感测图处理系统	DVP(Digital Video Plotter)
加拿大发射的雷达卫星	RADARSAT
加拿大卡尔加里大学研制的 GPS 数据处理软件	Kingspad
加拿大诺瓦泰	Novatel
加拿大诺瓦泰公司研制的 GPS 数据后处理软件	Waypoint
加权平均值	weighted mean value, weighted average value
加权融合	weight fusion
加速度	acceleration
加速计	accelerometers
假彩色	false color
假彩色合成	false color composite
假彩色摄影	false color photography

假彩色图像 false color image
假定,假设 assumption
假定平均数 assumed mean
假定值 assumed value
假定坐标系 assumed coordinate system
假分数 improper fraction
假设 hypothesis
假设,假定 postulate
假设检验 hypothesis testing
假设检验理论 theory of testing hypotheses
假想的力,伪力 fictitious force
间断(的) discontinuous
间隔,间距 spacing
间接法纠正 indirect scheme of digital rectification
间接平差 indirection adjustment
间接照明 indirect illumination
间曲线 half-interval contour
间曲线,半距等高线,补充等高线 auxillary contour
间曲线,半距等高线,补充等高线 supplementary contour
间隙率 gap probability
监测,监视,监视器,监测器 monitor
监测标志 monitoring markers
监测技术 monitoring technique
监测网 monitoring network

监测仪器	monitoring equipment
监测站	monitoring station
监督分类	supervised classification
监控站	monitor station
监理工程师	supervising engineer
监视台,检查台	check station
减法	subtraction
减量	decrement
减色印刷	reducing color printing
减速度	decelaration
减压电阻器,降压电阻器	dropping resistors
剪辑	clipping
剪切变形	shearing deformation
剪切力	shear
检测	detection
检测模板	detection template
检查板	check board
检查测深线	check line of sounding
检查线透写图	tracing of check line
检校,校准,标定	calibration
检修井	inspection manhole
检修孔	inspection hole
检验标准	test criterion
简单布格重力异常	simple Bouguer gravity anomaly
简单迭代法	simple iteration method
简谐运动	simple harmonic motion
简写符号	abbreviation

简易公路	hasty road
碱性电池	alkaline battery
建模	model building, modeling
建设工程规划许可证	building permit, land use permit, permission notes foe land use
建筑材料	construction material
建筑地图	construction map
建筑费用	building cost
建筑工程	building engineering
建筑工程测量	building engineering survey
建筑工地	building site
建筑规模	building size
建筑红线	property line
建筑红线,建筑界限	building line(BL)
建筑红线测量	property line survey
建筑立面	building facade
建筑密度	building coverage, building density
建筑面积	building area
建筑平面图	architectural plan, construction plan
建筑设计	building design
建筑摄影测量	architectural photogrammetry
建筑条例	building law
建筑物沉降观测	building subsidence observation
建筑物沉陷	settlement of buildings
建筑物间距	spacing of buildings
建筑物检测	building detection
建筑物提取	building extraction

J

建筑物稳定性 stability of structures
建筑物阴影 construction shadow
建筑物重建 building reconstruction
建筑现场,工地 project site
建筑线,红线 net line
建筑用地 land for building
建筑制图 architectural drafting
建筑轴线测量 building axis survey
建筑总面积 gross floor area
渐近误差 asymptotic error
渐进式图像检索 progressive image retrieval
降低成本 cost reduction
降级锥曲线 degenerated conic section
降维 reduction dimension
降雨量 amount of rainfall
交叉点平差 crossover adjustment
交叉定标 cross calibration
交叉分区索引 cross-tile indexing
交叉辐射 cross radiance
交叉口 junction
交叉路 crossing
交叉耦合效应 cross-coupling effect
交叉网线 cross-ruling
交错采样 interlaced sampling
交点 point of intersection
交互式的 interactive
交换格式 transfer format
交会法 intersection

J

交会角　intersection angle
交流电　alternating current
交通道　traffic lane
交通分配　traffic assignment
交通量　traffic volume
交通事故　traffic accident
交线条件　condition of intersection, czapski condition, scheimpflug condition
交线　line of intersection
交向摄影　convergent photography
胶片平整度　film flatness
胶印　offset printing
椒盐噪声　impulse noise
焦距　focal length
焦距锁定　focus lock
角点检测　corner detector / detection
角定向　angular orientation
角动量　angular momentum
角度闭合差　angular closure, misclosure of angles
角度测设　layout of an angle
角度方向二阶矩　angular second moment(ASM)
角度观测　angular observation
角度归算,方向改化　angular reduction
角度畸变　angular distortion
角度计算　angle calculation
角度检定　angular calibration
角度精度　accuracy of angle

角度距离交会	angular or distance intersection
角度平差	adjustment of angles
角度误差	angular error
角分辨率	angular resolution
角速度	angular velocity
角锥体法	pyramid principle
脚手架,工作架	falsework
教育设施	educational facilities
校正曲线	adjustment curve
校准精度	calibration accuracy
校准器,检定器	calibrator
接边地图	adjacent map
接边纠正	overlapping area correction
接触晒印	contact printing
接触网屏	contact screen
接收机	receiver
接收机天线	antenna,receiver antenna
接收机阵列	antenna array,receiver array
接收机自主完好性监测	receiver autonomous integrity monitoring(RAIM)
接收中心	receiving center
接受区域	region of acceptance
接头、插头、转接器	connector
结点平差	adjustment by method of junction point
结构分析	structure analysis
结构化选取	structured selection
结构算法	structure algorithm

J

结构特征	object structure
结构元	structural element
结构状态智能模拟	intelligent simulation of structure state
结论	conclusion
截断误差	truncation error
截距	intercept
截面差改正	reduction to geodesic
截面积	cross section area
截尾	cuts off
截止高度角	cutoff angle, mask angle
解	solution
解缠	unwrap
解调	demodulation
解集	solution set
解析测图	analytical mapping
解析测图仪	analytical plotter
解析定向	analytical orientation
解析几何	analytical geometry
解析纠正	analytical rectification
解析空中三角测量	analytical aerial triangulation
解析空中三角测量	analytical aerotriangulation
解析铅垂线检校法	analytical plumb-line calibration
解析摄影测量	analytical photogrammetry
解析图根点	analytical mapping control point
介电常数	dielectric constant
界标	boundary demareation
界面接口选项	interface option

J

界石 boundary stone
界线 boundary line
界址点 boundary mark, boundary point
紧急报警系统 emergency alarm system
紧急出口,安全出口 emergency exit
紧急出口门 emergency exit door
紧急频道 emergency channel
紧急事故监察及支援中心 emergency monitoring and support centre
紧急事故控制中心 emergency control centre
紧急通道 emergency access
紧急制动器 emergency brake
近程定位系统 short-range positioning system
近地表测量 ground measurement
近地点 perigee
近海测量 offshore survey
近海航行图 offshore sailing chart
近红外 near infrared
近景摄影测量 close-range photogrammetry
近期建设规划 immediate plan
近日点 perihelion
近似垂直摄影 near vertical photography
近似等高线 approximate contour
近似地形面 telluroid
近似高度 approximate altitude
近似公式 approximate formula
近似计算 approximate calculation
近似平差 approximate adjustment

近似相对熵	approximate entropy
近似值	approximate value, approximation
近似坐标	approximate coordinates
进动	precessional motion
进动方向	direction of precession
进动角速度	precession angular velocity
禁航区	prohibited area
禁用调和函数	forbidden harmonics
经差	difference of longitude
经度	longitude
经度测定	determination of longitude
经度起算点	origin of longitude
经度圈	longitude circle
经度误差	error in longitude
经济地图	economic map
经济地位,经济状态	economic status
经济规划	economic planning
经济技术指标	technical economic index
经济特区	special economic zone
经纬圈	declination circle
经纬仪	theodolite, transit, transit instrument
经纬轴	declination axis
经验定向,目视定向	empirical orientation
经验概率	empirical probability
经验模态分解	empirical mode decomposition (EMD)
经验线性定标	empirical linear calibration

精度 precision
精度分析 precision analysis
精度估计 precision estimation
精度检验 accuracy test
精度评估,精度检验 accuracy assessment
精度设计 accuracy design
精度因子 dilution of precision (DOP)
精度指标 index of precision
精化布格重力异常 refined Bouguer gravity anomaly
精纠正 accurate rectification
精码 precise Code(P Code)
精密测距 precise ranging
精密垂准 precise plumbing
精密单点定位 precise point positioning (PPP)
精密导线测量 precise traversing
精密定位服务 precise positioning service (PPS)
精密工程测量 precise engineering survey
精密工程控制网 precise engineering control network
精密三角测量 precise triangulation
精密水准测量 precise leveling
精密水准仪 precise level
精密星历 precise ephemeris
精密准直 precise alignment
精密准直测量 precise alignment observation
精确安置 fine setting
精确解 exact solution
精确量测 accurate measurement
精确值 exact value

J

精细农业,精细农作	precision agriculation (farming)
井探工程测量	shaft prospecting engineering survey
井下空硐测量	underground cavity survey
景观	landscape
景观特征	landscape pattern
景深	depth of field
景物反差	object contrast
景象匹配	scene matching
警示标志	caution sign
净空,净度	clearance
净空区测量	clearance survey
净面积	net area
径向基函数	radial basis function(RBF)
径向基函数神经网络	radial basis function neural networks (RBFN)
径向畸变	radial distortion
径向视差	radial parallax
静吃水	static draft
静电复印	xerography
静荷载	static load
静态传感器	static sensor
静态定位	static positioning
静止状态	static situation
镜面反射	mirror reflection, specular reflection
镜头	shot
镜头边缘检测	shot boundary detection
镜头畸变	lens distortion

J

镜向反射	specular reflection
纠正,整流	rectification
纠正图底	basis for rectification
纠正仪	rectifier, transformer
纠正元素	element of rectification
居民地	residential area
居民区	residential district
居住密度	living density, residential density
居住权	right of occupancy
居住使用面积	living floor space
局部规划	local planning
局部极大(值)	local maximum
局部极小(值)	local minimum
局部精度估计	local precision estimate
局部匹配	local matching
局部增强滤波	local enhancement and filtering
局部自动搜索	local automatic searching
局域差分 GPS	local differential GPS
局域增强系统	local area augmentation system (LAAS)
矩	moment
矩匹配	moment matching
矩形分布	rectangular distribution
矩形分幅	rectangular subdivision
矩阵	matrix
矩阵乘法	matrix multiplication
矩阵的迹	trace of a matrix
矩阵的阶	dimension of a matrix

J

矩阵的逆 inverse of a matrix
矩阵的秩 rank of a matrix
矩阵方程 matrix equation
矩阵广义逆 generalized inverse of a matrix
矩阵加法 matrix addition
矩阵运算 matrix operation
拒绝区域 region of rejection
距离 distance
距离分辨率 range resolution
距离公式 distance formula
距离估测 range estimation
距离计算 distance calculation
距离检核 distance check
距离均方根(误差) distance root mean square (error)
距离判决函数 distance decision function
距离匹配 distance-based matching
距离图像 range image
聚合(遥感尺度转换) scaling up
聚类 clustering
聚类分析 clustering analysis
聚类树 clustering tree
聚类统计 class statistic
卷积 convolution
决策 decision making
决策树 decision-making tree
绝对点位误差椭圆 absolute(point)error ellipses
绝对定位 absolute positioning
绝对定向 absolute orientation

J

绝对定向元素	elements of absolute orientation
绝对高程	absolute altitude
绝对航高	absolute flying height
绝对极大值	absolute maximum
绝对极小值	absolute minimum
绝对精度	absolute accuracy, absolute precision
绝对平均误差	absolute average error
绝对摄动	absolute perturbation
绝对视差	absolute parallax
绝对位置	absolute position
绝对温度	absolute temperature
绝对温度单位	Kelvin
绝对误差	absolute error
绝对盐度	absolute salinity
绝对阈	absolute threshold
绝对值	absolute value
绝对重力	absolute gravity
绝对重力测量	absolute gravity measurement
绝对重力仪	absolute gravimeter
绝对坐标	absolute coordinate
军用地图	military map
军用海图	military chart
均方根	root mean square (RMS)
均方根误差,中误差	root mean square error(RMSE)
均衡大地水准面	isostatic geoid
均衡椭球	equilibrium spheroid
均匀分布	uniformly distributed

均匀性、同性	homogeneity
均匀颜色空间	uniform color space
均值	mean value
均值漂移模型	mean shift model
竣工测量	as-built survey

K

开标	bid opening
开采沉陷观测	mining subsidence observation
开采沉陷图	mining subsidence map
开窗	windowing
开发区	development area
开放式地理数据互操作规范	Open Geodata Interoperability Specification
开放式地理信息系统	open geographic information system, Open GIS
开放式地理信息系统协会	Open GIS Consortium(OGC)
开放数据库互联	open database connectivity (ODBC)
开普勒行星运动定律	Kepler's Laws of Planetary Motion
开区间	open interval
开挖	excavation
开挖区,开挖面积	excavation area
开挖引导	excavation guidance
勘测	reconnaissance
勘测报告	reconnaissance report
勘测地球物理	exploration geophysics
勘测图	reconnaissance diagram
勘探基线	prospecting baseline
勘探网测设	prospecting network layout

勘探线测量	prospecting line survey
抗差估计,稳健估计	robust estimation
抗干扰	anti-jamming
抗震建筑	earthquake-proof construction
考古摄影测量	archaeological photogrammetry
考古学	archaeology
柯氏力	coriolis force
柯西-许瓦尔兹不等式	Cauchy-Schwarz inequality
柯西序列	Cauchy sequence
柯西主值	Cauchy principal value
科技馆	science museum
科技术语	scientific terminology
科学计数法	scientific notation
可变速度	variable velocity
可变速率	variable speed
可变性	variability
可持续发展	sustainable development
可持续性	sustainablility
可调谐滤光片	tunable filter
可航最浅水深	controlling depth
可积函数	integrable function
可加性	additive property
可见近红外	visible and near-infrared(VNIR)
可见性好	good sky visibility
可交换律	commutative law
可靠性	reliability
可靠性设计	reliability design

K

可扩展置标语言　extensible markup language(XML)
可逆的　invertible
可逆矩阵　vegular matrix
可逆性　invertibility
可逆性条件　invertibility condition
可视度函数　visibility function
可视化　visualization
可视化接口　visualization interface
可数集　countable set
可通行　right-of-way
可行解　feasible solution
可行性研究　feasibility study
可用性　availability
克　gramme(g)
克拉克椭球　Clarke spheroid
克拉索夫斯基椭球　Krasovsky ellipsoid
克莱罗定理　Clairaut theorem
克里金法　Kriging
克隆选择　clone selection
刻度盘　divided circle
刻画板　reading scale
刻绘　scribing
刻图仪　scriber
坑道平面图　adit planimetric map
坑探工程测量　adit prospecting engineering survey
空白图幅　blank sheet
空地定位　air-to-ground positioning
空盒气压计　aneroid barometer

空间波	space waves
空间布局	space planning
空间部分	space segment
空间参照系	spatial reference system
空间插值	spatial interpolation
空间查询	spatial query
空间大地测量技术	space geodetic techniques
空间大地测量学	space geodesy
空间单元	spatial unit
空间定向	spatial orientation
空间对象	spatial object
空间分辨率	spatial resolution
空间分布	spatial distribution
空间分析	spatial analysis
空间改正	free-air correction
空间关联规则	spatial association rule
空间关系	spatial relationship
空间后方交会	space resection
空间几何约束	geometric constraint in space
空间建模	spatial modeling
空间结构	space structure
空间聚类	spatial clustering
空间决策	spatial decision
空间科学	space science
空间隶属度	spatial fuzzy membership
空间模拟	spatial simulation
空间模型,立体模型	space model
空间频谱	spacial frequency

空间前方交会　space intersection
空间设计　space design
空间数据　spatial data
空间数据仓库　spatial data warehouse
空间数据基础设施　spatial data infrastructure(SDI)
空间数据交换网站　spatial data clearing house
空间数据结构　spatial data structure
空间数据库管理　spatial database management
空间数据库管理系统　spatial database management system
空间数据挖掘　spatial data mining
空间数据转换　spatial data transfer
空间数据转换标准　spatial data transfer standard (SDTS)
空间索引　spatial indexing
空间特征　spatial feature
空间拓扑关系　spatial topology
空间相关　spatial correlation
空间斜方位投影　space oblique azimuthal projection
空间信息科学　spatial information science
空间信息可视化　visualization of spatial information
空间信息网格　spatial information grid (SIG)
空间异常　free-air anomaly
空间直角坐标　spatial rectangular coordinate
空间重力加速度　gravity acceleration in space
空间属性　spatial attribute
空间资源储备　space resource storage
空间自相关　spatial autocorrelation

空间坐标	spatial coordinates
空运的,空降的,机载的	airborne
空中基线	air baseline
空中基线倾角	air base inclination
空中三角测量	aerial triangulation, aerotriangulation
空中三角网模型	aerial triangulation block
孔,(照相机)光圈,孔径	aperture
控制部分	control segment
控制测量	control survey
控制测量区	control area
控制点	control point
控制点坐标	control-point coordinate(s)
控制室	control room
控制网	control network
控制网分析	analysis of control networks
控制网精度	accuracy of network, control network density
控制网优化设计	optimal network design
控制网原点	origin of control network
控制站	control station
库容测量	reservoir survey
跨,跨度	span
跨河水准测量	river-crossing leveling
块状图	block diagram
快车道,高速公路	clear way

快速独立分量分析　fast independent component analysis
快速傅立叶变换　fast Fourier transform(FFT)
快速近似主成分分析算法　fast approximate principal component analysis algorithm
快速求解整周模糊度　fast integer ambiguity resolution
快速提升小波变换　fast lifting wavelet transform (FLWT)
宽波段　wide band
宽带多光谱空间　wide-band multi-spectral space
宽角数字相机　wide-angle digital camera
宽巷　widelane
矿产图　mineral deposits map
矿场平面图　mining yard plan
矿区控制测量　mining area control survey
矿山测量　mine survey
矿山测量图　mining map, mining surveying
矿体几何[学]　mineral deposits geometry
矿物开采　mineral extraction
框标　collimation mark, fiducial mark
框标点　collimation point
框幅摄影机　frame camera
框幅式　frame perspective
框架结构　frame construction, skeleton structure
框架小波变换　frame wavelet
扩建物　extended structure
扩散函数　spread function

扩散系数	diffusion coefficient
扩散转印	diffusion transfer
扩展分形	extended fractal

L

拉格朗日插值多项式	Lagrange interpolating polynomial
拉格朗日定理	Lagrange theorem
拉科斯特-隆贝格重力仪	Lacoste-Rormberg gravimeter
拉普拉斯点	Laplace point
拉普拉斯方位角	Laplace azimuth
拉普拉斯面调和函数,拉普拉斯面谐函数	Laplace surface harmonics
兰勃特投影	Lambert projection
蓝底图	blue print
缆,索	cable
劳动强度低	low labour intensity
雷达	radar
雷达测高仪	radar altimeter
雷达测角	radarclinometry
雷达覆盖区	radar overlay
雷达干涉测量	radar interferometry
雷达图像	radar imagery
雷达图像测量学,雷达图像测量技术	radargrammetry
雷达应答器	radar responder
雷达影像匹配	radar scene matching
雷达指向标	radar ramark

类别视觉感受	perceptual grouping
类内方差	variance within clusters
类型	category
类型地图	typal map
累积空间覆盖率	cumulative space covering ratio
累积误差	accumulation of error
棱镜	prism
冷启动	cold start
厘米	centimeter
离岸流	rip current, rip surf
离差	dispersion
离均差	deviation from the mean
离散	discrete
离散傅立叶变换	discrete Fourier transform(DFT)
离散均匀分布	discrete uniform distribution
离散控制系统	discrete control system
离散数据	discrete data
离散随机变量	discrete random variable
离散余弦变换	discrete cosine transform(DCT)
离心力	centrifugal force
离心力位	centrifugal potential
离心圆	eccentric circles
离子浓度	ion concentration
黎曼空间	Riemannian space
里程碑	kilometer stone
理论大地测量学	theoretical geodesy
理论地图学	theoretical cartography
理论概率	theoretical probability

理论基础	theoretical foundation
理论最低潮位面	theoretical lowest tide surface
力高	dynamic height
历史地图	historical map
历书,概略星历	almanac
历书,星历表	ephemeris
历书日	calendar day
历书时	ephemeris time
历元	epoch
历元平极	mean pole of epoch
立方根	cubic root
立方毫米	cubic millimeter (mm)
立方米	cubic meter (m)
立体补偿	stereo compensation
立体测量的	stereometric
立体测图	stereocompilation
立体测图仪	stereoplotter
立体的,立体观测的	stereoscopic
立体的,立体观测的,立体镜的	stereo
立体地,体视地	stereoscopically
立体地图	relief map
立体观测	stereoscopic observation
立体观察	stereo view
立体绘图	stereoplotting
立体镜	stereoscope
立体量测仪,视差量测仪,视差杆,体积计	stereometer

L

立体模型　stereoscopic model
立体判读仪　stereointerpretoscope
立体区域网平差,三维区域网平差　stereoblock adjustment
立体摄影测量　stereophotogrammetry
立体摄影机　stereocamera, stereometric camera
立体视觉　stereoscopic vision
立体视觉敏感度　stereoscopic (vision) acuity
立体卫星　stereosat
立体像对　stereopair, stereo image pair, stereo photopair
立体眼镜　stereo eyewear, stereo glasses
立体影像　stereo images
立体正射像片　orthophoto stereomate
立体自动测图仪　stereoautograph
立体坐标量测仪　stereocomparator
隶属度　membership
隶属函数　membership function
粒子加速器测量　particle accelerator survey
连接点　tie point
连通性　connectivity
连续调　continuous tone
连续定向　successive orientation
连续对比　successive contrast
连续跟踪　continuous tracking
连续观测　continuous observation
连续函数　continuous function
连续航带　continuous strip

连续数据	continuous data
连续性	continuity
连续运行参考站系统	continuously operating reference system(CORS)
帘幕快门	focal plane shutter,curtain shutter
联合分布	jiont distribution
联合分布函数	jiont distribution function
联合光束法平差	combined bundle block adjustment
联合平差	combined (simultaneous) adjustment
联合熵	union entropy
联机空中三角测量	on-line aerophotogrammetric triangulation
联立不等式	simultaneous inequalities
联立方程	simultaneous equations
联系测量	connection survey
联系三角形	connecting triangle
联系三角形法	connecting triangle method
链码	chain code
梁,桁条,(光线的)束	beam
梁,钢桁的支架	girder
两井定向	single wire method: two shafts
亮度	illuminating power,lightness
亮温	brightness temperature
亮温差值	brightness temperature difference
量测摄影机	metric camera
量程范围	measuring range

L

量化	quantization, quantizing
量化评价	quantitative evaluation
列矩阵,列向量	column matrix,column vector
列向量	column vector
劣质工程	poor engineering
裂缝观测	fissure observation
裂缝监测仪	crack monitoring pins
邻边	adjacent side
邻带方里网	grid of neighboring zone
邻角	adjacent angle
邻近效应	adjacent effect
邻图对接对比	comparison with adjacent chart
邻域	neighborhood
邻域匹配	block-matching
邻域识别	neighbor identification
邻元法	neighborhood method
林地	forest land
林业测量	forest survey
临边扫描	limb scanning
临界点	critical point
临界分析,门槛理论	threshold analysis
临界域	critical region
临界值	critical value
临时工程	temporary works
临时工程,临时建筑	temporary construction
临时海图	provisional chart
临时性建筑,临时性房屋	temporary building

L

灵敏度,感光度	sensitivity
灵敏度设计	sensitivity design
岭参数	ridge parameter
岭估计	ridge estimation
岭-压缩估计组合	ridge-stein combined estimator
菱形	rhombus
零点检查	zero check
零点漂移	zero drift
零点漂移改正	correction for zero drift
零点校正	zero calibration
零基线	zero baseline
零假设	null（zero）hypothesis
零阶模型	zero-order model
零矩阵	null matrix,zero matrix
零类设计:基准设计	zero-order design(ZOD): the datum problem
零米等深线	zero contour
零线改正	correction of zero line
零向量	null vector,zero vector
零因子	zero factor
零中频矢量滤波	zero intermediate frequency (ZIF) vector filtering
领海	territorial seas
领海基点	baseline point of territorial sea
领海基线	baseline of territorial sea
领域,领土	territory
浏览器	brower
流程图	flow chart

L

流动接收机	roving receiver
流动站	rover
流量	flow discharge
流速	current velocity
流向	stream direction
流域规划	watershed planning
六边形	hexagon
六分仪	sextant
露天采矿图	opencast mining plan
露天矿测量	opencast survey
陆表水指数	land surface water index
陆地表面温度	land surface temperature
陆地卫星(美国)	landsat
陆缘海	epicontinental sea
滤波	filtering
滤光镜	glass filter
滤光镜,滤光片,消色器	colour absorber
滤光片,滤光镜	filter
路标,方向标	guide post
路灯	lamp post
路肩,护坡道	berm
路肩,路边	road border
路径	path
路径查找	path finding
路径分析	route analysis
路线、路程、航线	course
旅游地图	tourist map

绿地率	greening rate
绿地面积	green area
绿化城市	green-planted city
轮廓提取	contour extraction
轮廓追踪	contour tracing
罗德里格矩阵	Lodriguez matrix
罗兰-C 定位系统	Loran-C positioning system
罗兰导航系统	Long Range Navigation (LORAN)
罗兰海图	Loran chart
罗盘	compass
罗盘方位角	compass bearing
罗盘经纬仪	compass theodolite
罗盘误差	compass error
罗盘仪测量	compass survey
罗盘仪导线	compass traverse
罗盘仪导线网	network of compass traverse
逻辑兼容，逻辑一致性	logical consistency
逻辑斯蒂法	logistic model
逻辑推论	logical deduction
螺旋	screw
螺旋楼梯	spiral stairs
螺旋线	spiral
螺旋线几何特性	spiral curve geometry
落潮	ebb tide
落潮流	ebb current

M

马尔柯夫过程	Markov process
马尔柯夫链	Markov chain
马尔柯夫随机场	Markov random field(MRF)
马尔柯夫随机网	decomposable Markov network (DMN)
码分多址	code division multiple access (CODE)
码相关技术	code correlation technique
码相位	code phase
脉冲响应函数	impulse response fuction
满秩的,非奇异的	nonsingular
漫反射	diffuse reflectance
漫游	walkthrough
盲点消除	blind spot eliminating
盲色片	achromatic film
锚地测量	anchorage survey
锚地图	anchorage chart
锚固	rock bolting
卯酉面	prime vertical section
卯酉圈,东西圈	prime vertical circle
卯酉圈曲率半径	radius of curvature in prime vertical

美国/德国 Z/I Imageing 公司的卫星遥感测图处理系统	ImageStation
美国 Automatric 公司的卫星遥感测图处理系统	analytical image matching system (AIMS)
美国 GPS 精密定位与定轨软件	GAMIT
美国 JPL 研发的非差精密定位与定轨软件	GIPSY
美国 LH Systems 的卫星遥感测图处理系统	Digital Photogrammetry Workstation(DPW)
美国阿什泰克	Ashtech
美国的单面阵航空数码相机	EQ-90mm-CLR
美国的卫星遥感测图处理系统	I2S Digital Plotter
美国地质测量局	United States Geological Survey
美国国防测图局的卫星遥感测图处理系统	Digital Stereo Compatator/Compiler (DSCC)
美国海军海洋局	U. S. Naval Oceanographic Office
美国海军海洋学数字数据中心	Navy Fleet Numerical Oceanographic Center
美国和加拿大发射的合成孔径测试雷达	Sir-C
美国联邦地理数据委员会	Federal Geographic Data Committee,USA,FGDC
美国麦哲伦	Magellan

M

美国天宝	Trimble
门槛,临界值	threshold
蒙绘	tracing
蒙片	mask
蒙塞尔色系	Munsell color system
蒙特-卡罗方法	Monte_carlo method
米	meter
密度	density
密度分割	density slicing
密度分析	density analysis
密度计	densitometer
密度检测	density detection
密度流	density current
密码	cipher
幂等,等幂	idempotent
幂等阵	idempotent matrix
免疫算法	immune algorithms
面积采样	area sampling
面积测量	area survey
面积计算	area calculation
面目标	extended target
面向对象	object oriented
面状符号	area symbol
面状目标	area target
瞄直法	sighting line method
秒表	stop watch
民用建筑	civil architecture, civil building
民用日	civil day

M

民用时	civil time
明礁	bare rock
明亮色调,浅色调	light tone
明挖回填法	cut and cover method
明显地物点	outstanding point
模板	pattern plate
模板匹配	template matching
模糊 c-均值	fuzzy c-means(FCM)
模糊度	ambiguity, fuzzy degree
模糊度解算	ambiguity resolution
模糊度解算技术	ambiguity resolution technique
模糊对象	fuzzy objects
模糊分类	fuzzy classification
模糊分类法	fuzzy classifier method
模糊分析模型	fuzzy analysis model
模糊集	fuzzy set
模糊监督分类	fuzzy supervised classification
模糊紧支度	fuzzy compactness
模糊聚类	fuzzy clustering
模糊决策	fuzzy decision
模糊隶属度函数	fuzzy membership function
模糊逻辑	fuzzy logic
模糊容差	fuzzy tolerance
模糊神经网络	fuzzy neural network
模糊数学	fuzzy mathematics
模糊影像	fuzzy image
模拟	analogue
模拟(迭代)法	iterative (trial and error) approach

M

模拟,模仿,仿真	simulation
模拟磁带地震仪	analog seismometer
模拟地图	analog map
模拟加温-退火算法	simulated melt-annealing algorithm
模拟空中三角测量	analog aerotriangulation
模拟立体测图仪	analog stereoplotter
模拟摄影测量	analog photogrammetry, analogue photogrammetry
模拟退火算法	simulated annealing arithmetic
模拟遥感数据	simulated remote sensing images
模式分析	pattern analysis
模式识别	pattern recognition
模式语言	pattern language
模数转换器	analogue/digital converter
模型底图	model base
模型点,空间点	model point
模型空间	model space
模型连接	bridging of model
模型融合	model combination
模型缩放	scaling of model
摩擦	friction
摩擦系数	friction coefficient
末速度	final velocity
莫尔条纹图,叠栅条纹图	Moiré topography
墨卡托投影	Mercator projection
母函数,生成函数	generating function
母集	super set

母线,生成元	generator
目标定位	object location, target location
目标反射器	target reflector
目标分辨率	object resolution
目标高	height of target(HT)
目标函数	objective function
目标检测	target detection
目标区域	object region
目标识别	object identification, object recognition
目标提取	targets extraction
目视判读	visual interpretation
目视天顶仪	visual zenith telescope

N

纳米(毫微米)10^{-9}米	nanometer
南半球	south hemisphere
南黄极	south ecliptic pole
南极	south pole
南极科学考察委员会	Scientific Committee on Antarctic Research(SCAR)
南极圈	antarctic circle
南极研究专设委员会	Special International Committee on Antarctic Research
南极制图	surveying and mapping of Antarctica
南天极	south celestial pole
挠度	flexivity
挠度观测	deflection observation
内摆线	hypocycloid
内波	internal wave
内部定向	interior orientation
内部精度	internal accuracy
内部可靠性	internal reliability
内部通信联络系统，联络用对讲电话装置	intercom
内插点	interpolation point
内存	internal memory
内定向	internal orientation

内方位元素	data of inner orientation
内核	inner core
内积	inner product
内角	interior angle
内接三角形	inscribed triangle
内力	internal force
内陆海	enclosed sea
内能	internal energy
内切圆	inscribed circle
内约束	inner constraints
能见敏锐度	visibility acuity
能量边缘	energy edge
拟合良好性检验,适合度检测	goodness-of-fit test
拟稳平差	quasi-stable adjustment
逆的	inverse
逆反摄影测量	inverse photogrammetry
逆关系	inverse relation
逆矩阵	inverse matrix
逆时针	counterclockwise
逆时针方向	anti-clockwise direction, counter-clockwise direction
逆算问题	inverse problem
逆向断层	reverse fault
逆向反射器,反光镜	retroreflectors
逆向辐射温度	reverse radiative temperature
逆转点	reversal or turning point
逆转点法	reversal points method, turning point (or reversal) method

N

年降雨量	annual rainfall
年平均海面	annual mean sea level
鸟瞰图	bird's eye view map
牛顿万有引力定律	Newton's law of universal gravitation
扭曲	twist
农业气象模型	agro-meteorological model
农作物单产	crop yield prediction

O

欧几里得几何	Euclidean geometry
欧几里得空间	Euclidean space
欧几里得算法	Euclidean algorithm
欧空局	European Space Agency(ESA)
欧拉公式	Euler's formula
欧洲静地导航覆盖服务	European Geostationary Navigation Overlay Service(EGNOS)
欧洲遥感卫星	ERS(Europe remote sensing satellite)
偶函数	even function
偶极子换能器	doublet transducer
偶然误差	accidental error, random error
偶数	even number

P

排列	permutation
排水工程	water drainage works
排水管	drain pipe, outlet pipe
排水设计	drainage design
排水系统	draignage system
盘右	face right
盘左	face left
判别分析	discriminatory analysis
判别式	discriminant
判定框	decision box
判读,判释,解释	interpretation
判读,识别	indentification
判读要素	interpretation element
判读仪	interpretoscope
旁向倾角,横摇	lateral tilt, roll
旁向重叠	lateral overlap, side overlap, side lap
抛射运动	projectile motion
抛物面	paraboloid
抛物曲线	parabolic curves
抛物线	parabola
配置	collocation
膨胀	dilatation
膨胀,涌浪	swell

P

劈窗算法	split-window algorithm
皮米(微微米)10^{-12}米	picometer
匹配窗口	match window
匹配滤波	matched filter
偏差	bias, deviation
偏角,磁偏角	angle of deviation
偏角法	deflection angle, method of deflection angle
偏斜	skewness
偏斜分布	skew distribution
偏斜线	skew line
偏心常数	excentricity
偏心改正	eccentricity correction
偏心畸变差	decentering distortion
偏心角	eccentric angle
偏心率,离心率	eccentricity
偏移量	offset
偏振反射	polarized reflection
偏振光	polarized light
偏振光立体观察	vectograph method of stereoscopic viewing
偏自相关函数	partial autocorrelation function
偏最小二乘回归	partial least squares regression
拼接缝消除	seam line removal
频分多址	frequency division multiple access (FDMA)
频率	frequency

P

频率响应函数 frequency response function
频偏 frequency offset
频漂 frequency drift
频谱,波谱 spectrum
频数分布,频率分布 frequency distribution
频域 frequency domain
平差,校正,调整 adjustment
平差法 adjustment method
平差高程 adjusted elevation
平差计算 adjustment computation
平差角 adjusted angle
平差量 adjusted quantity
平差值 adjusted value
平潮 slack tide
平春分点 mean vernal equinox
平方,正方形 square
平方根 square root
平方毫米 squaremillimeter
平方米 square meter
平分 bisection
平分线,等分线 bisector
平恒星时 mean sidereal time
平衡 equilibrium
平衡潮 equilibrium tide
平衡温度 equilibrium temperature
平滑 smoothing
平滑度 smoothness
平滑技术 smoothing techniques

平极　mean pole
平经度,平黄经　mean longitude
平距　horizontal distance
平距成像　horizon-range projective imaging
平均半潮面　mean half-tide level
平均潮汐面　mean tide level
平均吃水　average draft
平均大潮差　mean spring range
平均大潮低潮面　mean low water springs
平均大潮低低潮面　mean lower low water springs
平均低潮　mean low water
平均低潮间隙　mean low water interval
平均低低潮面　mean lower low water
平均方位角　mean azimuth
平均高潮　mean high water
平均高潮间隙　mean high water interval
平均高高潮　mean higher high water
平均海面　mean sea surface
平均海面归算　seasonal correction of mean sea level
平均海面季节性改正　seasonal correction of sea level
平均海水面　mean sea level(MSL)
平均剖面曲率　mean profile curvature
平均曲率半径　mean radius of curvature
平均视差　mean parallax
平均水面　mean water level
平均速率　average speed
平均误差　average error

P

平均小潮差	mean neap range
平均正常重力	mean normal gravity
平均值	average value
平均重力	mean gravity
平流参数	advection parameter
平流雾	advection fog
平面	horizontal plane, plane
平面布置图	layout plan
平面测量,平面测量学	plane surveying
平面场定标	flat calibration
平面定线	horizontal alignment
平面方程	simultaneous equations of planes
平面格网	planar grid
平面控制点	horizontal control point, horizontal control station
平面控制精度	accuracy for horizontal control
平面控制网,水平控制网	horizontal control network
平面曲线测设	horizontal curve layout, plane curve location
平面三角形	plane triangle
平面图,鸟瞰图	plan view, planimetric map
平面图形	plane figure
平面直角坐标	rectangular plane coordinate
平面直角坐标,格网坐标	grid coordinate
平面坐标	horizontal coordinate, plane coordinate

P

平时	mean time
平世界时	universal time-two (UT2)
平台	platen
平太阳日	mean solar day
平太阳时	mean solar time
平坦地	level terrain
平纬圈,等高圈	almucantar
平稳	stationary
平稳随机过程	stationary stochastic process
平稳随机模型	stationary stochastic model
平稳性	stationarity
平稳性条件	stationarity conditions
平稳序列	stationary series
平稳值	stationary value
平行(直线)	parallel lines
平行力	parallel force
平行六面体	parallelepiped
平行圈半径	radius of parallel
平行圈曲率	curvature of parallel
平行射线近似	parallel ray approximation
平行四边形	parallelogram
平移	translation
平移参数	translation parameters
平子午线	mean meridian
评价体系	system of assessment
屏幕	screen
屏幕地图	screen map
坡度	slope

坡度,地面标高,室外地坪	grade
坡度测设	grade location
剖面	sectional plane
剖面图,断面图	sectional drawing
普通地图集	general atlas
普通海图	general chart
普通天文学	general astronomy
谱分析	spectral analysis
谱估计	estimation of the spectrum
谱间结构	spectrum structure
谱密度函数	spectral density function
谱相关	spectral correlation
曝光	exposure
曝光站,摄站	exposure station

Q

七边形	heptagon
期望	expectation
期望航线(从起点到终点的路线)	desired track (DTK)
齐次方程	homogeneous equation
奇异的	singular
奇异矩阵	singular matrix
奇异性,异常	singularity
奇异值分解	singular value decomposition (SVD)
起始方位角	initial azimuth
气候变化	climate change
气泡改正	bubble correction
气泡居中	centering of bubble
气象浮标	meteorological buoy
气象观测	moteorogical onservation
气象雷达	meteorological radar
气象卫星	meteorological satellite
气象灾害	weather disaster
气压高程测量,气压测高	barometric leveling
气压高程计,气压测高仪,高差仪	pressure altimeter
气压计	barograph, barometer

铅垂线　plumb line, vertical line
铅垂线轨迹法　vertical line locus(VLL)
铅直度控制　verticality control
前尺　fore rod
前方交会　intersection
前视　foresight(FS)
前向反射　face reflectance
前置放大器　pre-amplifier
浅地层地质剖面仪　shallow geological profile chart
浅地层剖面仪　sub-bottom profiler
浅色调　tint
浅滩　shoal
欠采样　undersampled
嵌入式隐马尔柯夫模型　embedded hidden Markov model (EHMM)
强度　intensity
强度值,亮度值　intensity value
强制对中　forced centering
桥墩　pier
桥墩定位　location of pier
桥墩施工　pier construction
桥墩中心　pier centre
桥梁测量　bridge survey
桥梁施工测量　bridge construction survey
桥梁轴线测设　bridge axis location
桥面　bridge deck
桥塔　pylon
桥台,桥墩　abutment

切点　point of tangency, tangent point
切距　tangent distance
切线支距法　tangent offset method
切向畸变　tangential distortion, tangential lens distortion
切向加速度　tangential acceleration
切向视差　tangential parallax
轻轨交通　light rail transit
倾角　inclination angle, tilt angle
倾角测定　determination of tilt
倾角方向　direction of tilt
倾角改正　correction for tilt angle, tilt adjustment
倾斜,倾角,尖端　tip
倾斜,斜角　inclination, tilt
倾斜的　oblique
倾斜观测　oblique observation, tilt observation
倾斜航空摄影像片　oblique aerial photograph
倾斜距离校正　slant range correction
倾斜流　slope current
倾斜摄影　oblique photography
倾斜摄影像片　oblique photograph
倾斜仪　clinometer, inclinometer
清绘　fair drawing
丘陵地　rolling terrain
求和法　summation
求和公式　summation formula
求积仪　planimeter

求积仪法	planimeter method
球带调和函数，球带谐函数	zonal spherical harmonics
球概率误差	spherical error probable
球面调和函数，球面谐函数	surface spherical harmonics
球面方位角	reference azimuth
球面三角形	spherical triangle
球面天文学	spherical astronomy
球面投影	stereographic projection
球面直角三角形	right spherical triangle
球面坐标	spherical coordinate
球体，椭球体	spheroid
球体谐函数	solid spherical harmonics
球谐函数	spherical harmonics
球心投影，极平投影	gnomonic projection
球形	sphere
区划地图	regionalization map
区间	interval
区间估计	interval estimate, interval estimation
区域	region
区域导航卫星系统	local navigation satellite system
区域地图集	regional atlas
区域地质调查	regional geological survey
区域地质图	regional geological map
区域分析	regional analysis
区域规划	regional planning

Q

区域海道测量委员会 Regional Hydrographic Commission
区域合并 region merging
区域控制 zone control
区域生长 region-growing
区域填充 region filling
区域网平差 block adjustment
区域增长算法 region growth
曲率 curvature
曲率半径 curvature radius
曲率改正 curvature correction
曲面 curved surface
曲面面积 curved surface area
曲面拟合 curved surface fitting
曲线 curve
曲线测设 layout of curves
曲线长度 length of curve
曲线的法线 normal to curve
曲线放样 curve layout
曲线光滑 line smoothing
曲线描绘(法) curve sketching
曲线描迹(法) curve tracing
曲线拟合(近似) curve fitting (approximation)
曲线隧道 curved tunnel
曲中点 midpoint of curve
去噪 denoising
全等 congruence
全等三角形 congruent triangles
全等图形 congruent figures

Q

全国地理信息标准化技术委员会	national technical committee for geographical information, CSBTS/TC 230
全极化	fully-polarized
全景的	panoramic
全景畸变	panoramic distortion
全景摄影	panoramic photography
全景摄影机	panoramic camera
全景图	overall perspective
全局优化	global optimization
全立体覆盖	full stereoscopic coverage
全能经纬仪,通用经纬仪	universal theodolite
全球板块运动	global plate motions
全球导航卫星系统	global navigation satellite system (GNSS)
全球电离层图	global ionosphere map(GIM)
全球定位系统	global positioning system(GPS)
全球轨道导航卫星系统(俄罗斯)	global orbiting navigation satellite system (GLONASS)
全球海平面	global mean sea surface
全球重力异常	global gravity anomalies
全日潮流	diurnal tidal current
全色的	panchromatic
全色红外片	panchromatic infrared film
全色片	panchromatic film
全色影像	panchromatic image
全数字化测图	fully digital mapping

全数字化自动测图系统(第一套数字摄影测量系统,19 世纪 60 年代在美国生产)	Digital Automatic Map Compilation System(DAMCS)
全天候	any weather condition
全天候测距仪	all weather terrestrial rangefinder
全息摄影	hologram photography
全息摄影,全息技术	holography
全息摄影测量	hologrammetry
全向的,无定向的	omnidirectional
全向天线	omnidirectional antenna
全站仪	total station
全振动	complete oscillation
全帧帧传输	full frame transfer
全自动	fully automated
全组合测角法	method in all combinations
权	weight
权函数	weight function
权矩阵	weight matrix
权逆阵	inverse of weight matrix
权系数	weight coefficient
确定性函数	deterministic function
群落结构	community structure
群速	group velocity
群样本检验	cluster sample

R

热点效应	hotspot effect
热辐射	heat radiation, thermal radiation
热辐射方向性	directional thermal emission
热惯量	thermal inertia
热红外	thermal infrared
热红外多光谱	thermal infrared multispectral scanner(TIMS)
热红外扫描仪	thermal infrared scanner
热红外探测器	thermal infrared detector
热红外图像	thermal infrared imagery, thermal IR imagery
热红外遥感	thermal infrared remote sensing, thermal IR remote sensing
热搅动,热激发	thermal agitation
热像仪	thermal imager
热源	thermal source
人工标志点	artificial target
人工地物	man-made objects
人工地物,人文要素	cultural feature
人工读数	manual reading
人工环境	man-made environment
人工建筑物	man-made structure
人工蚂蚁	artificial ants
人工免疫	artificial immune

人工免疫系统　artificial immune system
人工神经网络　artificial neural network(ANN)
人工智能　artificial intelligence(AI)
人机交互处理　interactive processing
人孔，检修孔　manhole
人口地图　population map
人口分布　population distribution
人口结构　population structure
人口密度　population density
人口统计数据　demographic data
人口统计数据库　demographic database
人口增长率　population growth rate
人口总密度　gross population density
人类活动　man-made activities
人力、物力和财力　human resources, material resources and cost
人为误差　personal error
人为影响　anthropogenic influence
人为原因　anthropogenic causes
人为灾害　man-made hazards
人行横道　pedestian crossing
人行天桥　over crossing, pedestian bridge
人眼视觉特性　human visual system(HVS)
人造地球卫星　artificial earth satellite
人造卫星　artificial satellite
刃首　cutting head
认知模式　cognitive pattern
认知制图　cognitive mapping

R

任意比例尺　arbitrary scale
任意常数　arbitrary constant
任意投影　arbitrary projection
任意原点　arbitrary origin
日本地球资源卫星合成孔径雷达　JERS-1
日本发射的地球资源卫星　JARS
日本索佳　Sokkia
日本拓普康　Topcon
日变　diurnal variation
日不等　diurnal inequality
日潮港　diurnal tide harbor
日平均海面　daily mean sea level
日蚀，月蚀，蒙蔽　eclipse
日心坐标系　Heliocentric coordinate systems
日月摄动　lunisolar perturbation
日月岁差　lunisolar precession
日月引力摄动　lunisolar gravitational perturbation
日照标准　insolation standard
日照计算　solar calculation
容积率　floor area ration, plot ration
容许荷载　permissible load
容许施工限差　construction tolerance
容许误差　allowable error
容许值　allowable value
融合　fusion
冗余度　redundancy

冗余码	redundant code
冗余信息	redundant information
入口	portal
入射	incidence
入射角	incident angle
软对应	softassign
软拷贝	soft copy
软拷贝摄影测量	softcopy photogrammetry
软门限滤波	soft thresholding filter
锐角	acute angle
锐角三角形	acute angle triangle
瑞利反射率	Rayleigh reflectance
瑞利散射	Rayleigh scattering
瑞士 Leica 公司和美国 GSI 公司联合推出的视频摄影测量系统	V-STARS
瑞士 Zurich 大学研制的 DTM 软件	CIP
瑞士伯尔尼大学研发的精密定位与定规软件	Bernese
瑞士的实时摄影测量系统	RTP
瑞士徕卡	Leica

R

S

赛博空间	cyberspace
三倍	triple
三倍角	triple angle
三边测量	trilateration
三边网	trilateration network
三差	triple-difference
三差相位观测	triple difference phase observation
三次方	cubic
三次方程	cubic equation
三等分	trisect
三焦张量	trifocal tensor
三角比	trigonometric ratio
三角闭合差	closing error of trangle
三角不等式	triangle inequality
三角测量	triangulation
三角测量基线	triangulation base
三角测量控制	trigonometric control
三角点	triangulation point
三角方程	trigonometric equation
三角高程测量	trigonometric leveling
三角高程网	trigonometric leveling network
三角函数	trigonometric function
三角函数表	trigonometric table
三角恒等式	trigonometric identity

S

三角基座	tribrach
三角矩阵	triangular matrix
三角锁	triangulation chain
三角锁网	chain network
三角网	triangulation network
三角形	triangle
三角形法	triangle method
三角形内角和	angle sum of a triangle
三角学	trigonometry
三脚架	tripod
三脚架基座	tripod base
三类设计:网的改造	third-order design (THOD): the improvement problem
三频相位模糊度解算	three-carrier ambiguity resolution (TCAR)
三丝水准测量	three-wire leveling
三维城市建模	three-dimensional city modeling
三维大地测量学	three-dimensional geodesy
三维地景仿真	three-dimensional terrain simulation
三维空间	three-dimensional space
三维量测鼠标	TopoMouse
三维模拟	three-dimensional simulation
三维网	three-dimensional network
三维显示	three-dimensional display
三维信息	three-dimensional information
三维坐标,空间坐标,立体坐标	three dimensional coordinate

三线阵　three-line array
三线阵 CCD　three line CCD scanners
三线阵 CCD 影像　three-line array CCD imagery
三线阵扫描仪　three line scanner
三项式　trinomial
三重积　triple product
散射测量　scatterometry
散射-辐射计　scattero radiometer
散射计　scatterometer
散射计辐射计组合　radiometer scatterometer
散射特性　scattering characteristic
扫测盲区　sweep blind zone
扫海　sweeping
扫海测量　sweep survey, wire drag survey
扫海具　sweeper
扫海区　sweep area
扫海深度　sweeping depth
扫描　scanning
扫描侧视角　side look angel
扫描次序　scanning order
扫描带,带状　strip
扫描带宽　swath width
扫描电子显微镜　scanning electron microscope
扫描轨迹　scanning trace
扫描角　scan angle
扫描镜　scanning mirror
扫描雷达成像技术　SCANSAR
扫描频率　scanning frequency

扫描数字化 scan-digitizing
扫描行 scan lines
扫描仪 scanner
色彩管理系统 color management system
色彩匹配 color matching
色彩平衡 radiometric alteration
色彩信号复现 reapparition of color signal
色彩信号提取 extraction of color signal
色调 hue, tone
色调调整 tone(response) adjustment
色度 chroma
色环 color wheel
森逊法则 Simpson's rule
沙尘监测 dust detection
沙尘源区 dust source area
沙盘地图 sand map
山地,丘陵地 upland
山谷 hollow
山谷线 valley line
山脊 crest
山脊线 crest line, ridge line
山腰,山坡 hillside
闪闭法立体观察 blinking method of stereoscopic viewing
扇球谐函数,扇球调和函数 sectorial spherical harmonics
扇式 sector
扇形角 angle of the sector

熵　entropy
熵编码　entropy coding
上部结构,上层建筑　superstructure
上界　upper bound
上邻游程　up neighbor run-length
上坡　ascending grade, upgrade
上三角阵　upper triangular matrix
上下视差　vertical parallax, y-parallax
上限　upper limit
上中天　upper transit
设计参数　design parameter
设计草图　sketch design
设计概算　preliminary estimate
设计高程　design elevation
设计贯通精度　designed breakthrough accuracy
设计阶段　design phase
设计矩阵　design matrix
设计年限　period of design
设计室,制图室　drawing office
设计水位　design water level
设计图纸　design paper
社会控制　society control
射电卫星　radio satellite
射电卫星空间网　net of radio satellites
射频　radio frequency (RF)
摄动　perturbation
摄谱仪　spectrograph
摄影,像片　photograph

摄影比例尺	photographic scale
摄影测量机	photogrammetron
摄影测量机器人	photogrammetric robot
摄影测量基线	photogrammetry baseline
摄影测量畸变差	photogrammetric distortion
摄影测量内插	photogrammetric interpolation
摄影测量学	photogrammetry
摄影测量仪器	photogrammetric instrument
摄影测量员	photogrammetrist
摄影测量坐标系	photogrammetric coordinate system
摄影处理	photographic processing
摄影的	photographic
摄影分区	flight block
摄影航线	flight line of aerial photography
摄影机检校	camera calibration
摄影机平台	camera platform
摄影机主距	principal distance of camera
摄影基线	air base, photographic baseline
摄影经纬仪	photo theodolite, camera transit
摄影天体测量学	photographic astrometry
摄影学	photography
摄影质量	photographic quality
摄影主距	photographic principal distance
摄站	camera station, exposure station
伸缩仪	extensometer
伸张,展	elongation
深度	depth
深度比例尺	scale of depth

深度改正	correction of soundings
深度感	depth perception
深度基准	sounding datum
深度基准面	depth datum
深度基准面保证率	guaranteed efficiency depth datum
深度计	bathymeter
深度索	depth wire
深度透写图	depth tracing
深海高原	submarine plateau
深海潜水器	bathyscaph
深海丘陵	abyssal hill
深空探测	deep space detection
深色调	shade
神经网络	neural network
神经元	nerve cell
甚长基线干涉测量	very long baseline interferometry (VLBI)
升	litre
升交点	ascending node
升交点黄经	longitude of ascending node
升交角距	argument of latitude
生活用水	living water use
生理视觉	biological vision
生态变化	ecological change
生态城市	ecological city
生态平衡	ecological balance
生态圈	ecosphere
生态系统	ecosystem

S

生态学 ecology
生物量 biomass
生物量指标变换 biomass index transformation
生物医学摄影测量 biomedical photogrammetry
声标 acoustic beacon
声波 sound wave
声波测深器,发声器,测深仪 sounder
声波束 sound wave beam
声波探测 depth sounding
声反射 sound reflection
声反射器 acoustic reflector
声呐 sonar
声呐方程 sonar equations
声呐浮标 sonobuoy
声呐目标 target of sonar
声呐图像镶嵌 sonar image mosaic
声呐扫海 sonar sweeping
声呐图像 sonar image
声散射 acoustic scattering
声速改正 sound velocity correction
声速计 velocimeter
声速平均值 average value of sound velocity
声速剖面 sound speed profile
声速梯度 sound velocity gradient
声速仪 sound velocity meter
声速跃层 transition layer of sound velocity
声图灰度变化 ultrasonogram gradient change

声图结构	ultrasonogram structure
声图判读	interpretation of echograms
声图图像	acoustic image
声线	sound ray
声线图	ray diagram
声相关计程仪	acoustic correlation log
声信号	acoustic signal
声学多普勒海流剖面仪	acoustic Doppler current profiler (ADCP)
声学浮标	sounding buoy
声学海流计	acoustic current meter
声学水位计	acoustic water level
声源	sound source
省际公路	interprovincial highway
剩余重力值	residual gravity
失锁	loss of lock
施工标准,建筑标准	standard of construction
施工测量	construction survey
施工队	construction crew
施工范围	works limit
施工方格网	square control network
施工放样	construction layout , setting-out survey
施工规范	constrction specification
施工监督	construction supervision
施工阶段	construction stage
施工控制网	construction control network
施工顺序	construction sequence

S

施工图	working drawing
施工详图	construction detail, detail work drawing
施工周期	construction cycle
湿地	wetland
湿度	humidity, moisture
湿度指数	moisture index
十米	decameter
十字丝	reticule
十字丝,十字瞄准线	crosshair, graticule
十字丝测微器	cross-wire micrometer
十字丝视差	parallax in cross-hairs, parallax in reticule
十字形水准器	cross level
石油勘探测量	petroleum exploration survey
时变	temporal variation
时变信息	spatio-temporal information
时变坐标	time-varying coordinate
时不变线性系统	time invariant linear system
时间差分	temporal differencing
时间分辨率	temporal resolution
时间精度因子	time dilution of precision(TDOP)
时间系统	time systems
时间序列分析	time series analysis
时间序列频域描述	frequency-domain descriptions of time series
时间序列时域描述	time-domain descriptions of time series

S

时间序列图 time series graphs
时空分布 spatial distribution and temporal variation
时空数据 spatio-temporal data
时空数据模型 spatial-temporal data model
时区 time zone
时圈 hour circle
时态特征 temporal characteristic
时域 time domain
时钟频率 clock frequency
时钟同步 clock synchronization
识别 recognition
识别码 identification code
识别效果 recognition effect
实部 real part
实测标,浮动测标 half-mark
实测原图 field map
实根 real root
实际距离 actual distance
实际面积,有效面积 actual area
实孔径雷达 real aperture radar
实孔径雷达图像 real aperture radar image
实码遗传优化算法 real coding genetic optimize algorithm
实时 real-time
实时差分 GPS real-time DGPS
实时差分改正 real-time differential correction
实时处理 real-time processing

实时动态定位　real time kinematic (RTK)
实时可视化　real-time visualization
实时摄影测量　real-time photogrammetry(RTP)
实时压缩　real-time compress
实时影像匹配　real image matching
实数　real number
实体　entity
实体对象　entity object
实体关系　entity relationship
实验场检校　test range calibration
实验概率　experimental probability
实验公式,经验公式　empirical formula
实验数据　empirical data
实用地图学　applied cartography
实用天文学　practical astronomy
实用盐标　practical salinity scale
实用盐度　practical salinity
实轴　real axis
矢量地图　vector map
矢量辐射传输理论　vector radiative transfer theory (VRT)
矢量跟踪　vector tracking
矢量化　vectorization
矢量绘图　vector plotting
矢量空间,向量空间　vector space
矢量量化　vector quantization
矢量数据　vector data
矢量数据结构　vector data structure

矢量-栅格转换	vector-to-raster conversion
使平滑,使光滑	smoothing adjustment
使用面积	floorage, usable area
使用年限	operational life, term of life
始点,起点	initial origin
世界大洋全图	world oceanic chart
世界地图集	world atlas
世界时	universal time (UT), universal time-zero(UT0)
世界时(加上极移改正)	universal time-one(UT1)
世界协调时	universal time coordinated(UTC)
市内交通	intracity traffic
市内交通,城市交通,市内运输	city traffic
市区	city proper, municipal area
市区中心	municipal center
市域	administrative region of a city
市域规划	city regional planning
市政当局,自治市	municipality
市政工程	public works
市政工程测量	public works engineering survey
市政厅	city hall
市政学	civics
示坡线	slope line
示误三角形	error triangle
势能	potential energy
势能场	potential field

S

视差	parallax
视差改正	correction for parallax
视场	field of view(FOV)
视场对比	simultaneous contrast
视场角效应	FOV effect
视地平线	apparent horizon
视方位角	apparent azimuth
视恒星时	apparent sidereal time
视角	viewing angles
视距	sighting distance, sight length
视距,视距尺	stadia
视距测量(法)	tacheometry, stadia
视距常数	stadia constant
视距乘常数	stadia multiplication constant
视距导线	stadia traverse
视距读数	stadia reading
视距间隔	stadia interval
视距丝,视距线	stadia hair
视觉变量	visual variable
视觉测量	visual measurement
视觉层次	visual hierarchy
视觉对比	visual contrast
视觉分辨敏锐度	resolution acuity
视觉分析	visual analysis
视觉跟踪	visual tracking
视觉景观	visual landscape
视觉立体地图	stereoscopic map
视觉平衡	visual balance

S

视觉信息,直观信息	visual information
视模型	perceived model
视频对象	video object
视频摄影测量	video photogrammetry
视频压缩	video compression
视区裁剪	view-frustum culling
视时	apparent time
视岁差,视进动	apparent precession
视太阳日	apparent solar day
视太阳时	apparent solar time
视图	view
视线	sight line
视线高	elevation of sight, height of sight, elevation of sight
视野	visual field
视准差	collimation error
视准线法	collimation line method
视准校正	collimation adjustment
视准轴	collimation axis
室内控制场	indoor control field
室内实验场	indoor test field
适应度	fitness
收方测量	wriggle survey
收敛	convergence
收敛的	convergent
收敛点	convergence point
收敛级数	convergent series
收敛角	angle of convergence

收敛率	rate of convergence
收敛区域	region of convergence
收缩	shrinkage
收缩膨胀因子	shrinkage and swell factors
手持水准仪	hand level
手调,人工调整	manual adjustment
手绘草图	free-hand sketch
首次回波	first echo (return)
首级控制网	primary control network
首级重力网	primary gravity network
首曲线,基本等高线	intermediate contour
首项	first term
首子午线	zero meridian
舒勒平均值	Schuler mean
输出框	output box
输出信号	output signal
输电线路测量	power transmission line survey
输入框	input box
输入信号	input signal
属性	attribute
属性表	attribute table
属性测度	attribute measure
属性查询	attribute query
属性精度	attribute accuracy
属性类型	attribute type
属性识别准则	attribute recognition criterion
属性数据	attribute data
属性数据文件	attribute data file

属性域	attribute domain
属性值	attribute value
树篱，障碍物	hedge
树形图	tree diagram
竖管，管体式水塔	standpipe
竖井	shaft, vertical shaft
竖井定向测量	shaft orientation survey
竖井激光指向[法]	laser guide of vertical shaft
竖盘指标差	index error of vertical circle
竖曲线	vertical curve
竖曲线半径	radius of vertical curve
竖曲线测设	vertical curve location
竖曲线几何形状	vertical curve geometry
竖向沉降	vertical settlement
竖向定线	vertical alignment
竖向净空	vertical clearance
竖向设计	vertical design
竖直摄影	vertical photography
竖轴，纵轴	vertical axis
数据保护	data protection
数据编辑	data editing
数据编码	data encoding
数据标准	data standard
数据采集	data capture
数据采集器	data collector
数据采样率	data sampling rate
数据仓库	data warehouse
数据处理	data processing

S

数据处理系统	data processing system
数据传输	data transmission
数据窗	data window
数据的可得性	data accessibility
数据分层	data layering
数据分发	data distribution
数据分类	data classification
数据分析与解释	data analysis and interpretation
数据格式	data format
数据更新	data revision, data update
数据共享	data sharing
数据管理	data management
数据归档	data archiving
数据获取	data acquisition
数据集	data set
数据集成	data integration
数据记录设备	datalogger
数据检索	data retrieval
数据简化	data simplification
数据交换格式	data exchange format(DXF)
数据交换中心	clearinghouse
数据结构	data structure
数据精度	data accuracy
数据可视化	data visualization
数据库	data base
数据库管理	database management
数据库管理系统	database management system
数据库管理员	database manager

S

数据库结构 database architecture, database structure
数据库设计 database design
数据库系统 database system
数据类型 data type
数据模型 data model
数据拟合 data fitting
数据清理 data cleaning
数据融合 data fusion
数据矢量化 data vectorization
数据搜集 data collection
数据缩减 data reduction
数据探测法 data snooping
数据通信串口协议 RS-232
数据挖掘 data mining
数据完全性 data completeness
数据完整性 data integrity
数据维护 data maintenance
数据下载 data download
数据现势性 data currency
数据相容性 data compatibility
数据压缩 data compression, data reduction, data compression
数据样品 data sampler
数据一致性 data consistency
数据源 data sources
数据真实性 data reality
数据正规化 data normalization

S

数据志 data lineage
数据质量 data quality
数据质量控制 data quality control
数据转换 data conversion, data transfer
数据字典 data dictionary
数控绘图机 digital plotter
数控绘图桌 digital tracing table
数量感 quantitative perception
数码城市 cybercity
数码相机 digital camera
数学 mathmatics
数学期望 mathematical expectation
数值积分法 numerical integration
数值精度 numerical accuracy
数值模拟 numerical simulation
数字 digit
数字表面模型 digital surface model(DSM)
数字测图 digital mapping
数字城市 digital city
数字磁带 digital tape
数字地理信息交换标准 digital geographic information exchange standard(DIGEST)
数字地面模型 digital terrain model(DTM), digital ground model (DGM)
数字地球 digital earth
数字地图 digital map
数字地图产品标准 standard of digital map product
数字地图学 digital cartography

数字地震仪　digital seismometer

数字读数　digital reading

数字高程模型　digital elevation model (DEM), digital height model (DHM), digital terrain elevation model (DTEM)

数字化　digitization

数字化地图　digitized map

数字化精度　digitizing accuracy

数字化文件　digital file

数字化仪　digitizer

数字化影像　digitized image

数字畸变模型　digital distortion model(DDM)

数字景观模型　digital landscape model(DLM)

数字纠正　digital rectification

数字立体摄影测量系统(瑞士 Kern 与英国剑桥 GEMS 公司共同研制,1988 年 7 月推出)　Digital Stereo Photogrammetric System(DSP1)

数字滤波　digital filtering

数字滤波器　digital filter

数字摄影测量　digital photogrammetry

数字摄影测量工作站　digital photogrammetric work station

数字数据网　digital data network(DDN)

数字双频回声测深仪　digital dual-frequency echosounder

数字水印　digital watermarking

数字图像处理　digital image processing

数字线画地图	digital line graph(DLG)
数字镶嵌	digital mosaic
数字信号源	digital signal generator
数字影像	digital image
数字栅格图	digital raster graphics(DRG)
数字栅格图	DRG
数字正射影像	digital orthoimage
数字正射影像图	digital orthophoto map (DOM)
数字制图数据标准	digital cartographic data standard
数字中国	digital China
数组,阵列	array
衰变常数,裂变常数	decay constant
衰变曲线,衰减曲线	decay curve
双边检验	two-sided test
双差	double-difference
双车道	double lane
双定向	double orientation
双介质摄影测量	two-medium photogrammetry
双立体同步摄影	dual stereo synchro photography
双目镜,双筒望远镜,双目的	binocular
双目立体视觉	stereo vision
双目序列影像	binocular sequence image
双频	dual-frequency
双频测深仪	dual-frequency sounder
双曲函数	hyperbolic function
双曲线	hyperbola
双曲线导航图	hyperbolic navigation chart

双曲线定位	hyperbolic positioning
双曲线定位系统	hyperbolic positioning system
双三次卷积	bi-cubic convolution
双色激光测距仪	two-color laser ranger
双筒天体摄影仪	double astrograph
双尾检验	two-tail test
双线偏振	dual polarization
双线性内插	bilinear interpolation
双向反射	bi directional reflectance
双向航道	two-way route
双向搜索	bi directional querying
双像素重采样	image resampling by pixel doubling
双主距	double principle distance
水尺零点	zero point of tide staff
水道测绘数据库	hydrographic surveying and charting database
水道学者，水道测量家	hydrographer
水库测量	reservoir survey
水库淹没线测设	setting-out of reservoir flooded line
水利工程测量	hydrographic engineering survey
水面	water surface
水泥	cement
水平测量，平面测量	horizontal survey
水平传感器	level sensor
水平度盘	horizontal circle
水平分量	horizontal component
水平渐近线	horizontal asymptote

S

水平角	horizontal angle
水平角闭合差	misclosure of horizontal angles
水平截面	horizontal section
水平精度	horizontal accuracy
水平精度因子	horizontal dilution of precision (HDOP)
水平射程	horizontal range
水平位移	horizontal displacement
水平位移观测	horizontal displacement observation
水平线	horizontal line
水平折光差	horizontal refraction error
水平轴,横轴,旋转轴	horizontal axis, transit axis
水汽反演	water vapor retrieval
水汽辐射仪	water vapor radiometer
水汽总量	columnar water vapour content
水色	color of water
水色计	color meter
水色遥感	ocean colour remote sensing, remote sensing of ocean color
水深测量	sounding
水深测量手簿	sounding field book
水深点间隔	sounding point interval
水生全息摄影系统	ultrasonic hologram system
水声定位	acoustic positioning
水声定位系统	acoustic positioning system
水声全息系统	acoustic holography system
水声通讯台	underwater sound communication station

水声学	underwater acoustics
水声应答器	acoustic responder, acoustic transponder
水体	water body
水体光谱	water spectrum
水体识别	water identification
水听器	hydrophone
水砣	lead
水砣测深	lead sounding
水位	water level
水位分带改正	correction of tidal zoning
水位改正	correction of water level
水位控制	control of water level
水位曲线	curve of water level
水位遥报仪	communication device of water level
水位站	gauge station
水温准确度	accuracy of sea water temperature
水文测量基准面	hydrographic datum
水文分析	hydrological analysis
水文观测,水文测验	hydrometry
水文特征	hydrologic patterns
水文要素	hydrologic features
水污染	water contamination
水系	water system
水系图	drainage map
水下电视	underwater television
水下激光测距仪	underwater laser range finder

S

水下立体摄影仪 underwater stereoscopic photographic apparatus
水下摄影测量 underwater photogrammetry
水下摄影机 underwater camera
水压测量 water pressure measurement
水压计程仪 pitometer log
水准闭合差 misclosure in leveling
水准测量 leveling
水准测量控制 leveling control
水准测量员 leveler
水准尺 leveling rod ,leveling staff
水准点标高 bench mark elevation
水准点之记 bench mark description
水准管轴 bubble axis,leveling bubble axis
水准环闭合差 misclosure of leveling loop
水准基点 benchmark(BM)
水准路线 leveling line
水准面 level surface
水准气泡偏差 bubble offset
水准器,气泡 bubble
水准椭球 level spheroid
水准网 leveling network
水准仪 level
水准原点 leveling origin
顺时针 clockwise
顺时针方向 clockwise direction
顺时针力矩 clockwise moment
瞬间地图 instantaneous map

S

瞬时海面　instantaneous sea level
瞬时极　instantaneous pole
瞬时加速度　instantaneous acceleration
瞬时视场　instantaneous field of view(IFOV)
瞬时速度　instantaneous velocity
瞬时速率　instantaneous speed
丝网印刷　silk-screen printing
斯托克斯理论　Stokes theory
撕膜片　peel-coat film
四边形，四边地　quadrilateral
四波束测深系统　four beam sounder
四面体　tetrahedron
四色印刷　four color printing
伺服马达　servo motor
似大地水准面　quasi-geoid
似大地水准面高　quasi-geoid height
似动力高　quasi-dynamic height
似然函数　likehood function
松弛算法　relaxation algorithm
搜索策略　search strategy
搜索区　searching area
素　prime
素数　prime number
算术　arithmetic
算术编码　arithmetic coding
算术平均　arithmetic mean
随机　random
随机(偶然)误差　accident error

随机变量	random variable
随机抽样	random sampling
随机过程	stochastic process
随机模型	stochastic model
随机试验	random experiment
随机数	random number
随机样本	random sample
岁差,进动	precession
碎部测量	detail survey
碎部点	detail point
隧道测量	tunnel survey
隧道导向系统	tunnel guidance system
隧道断面仪	tunnel profiler
隧道盾构法	tunnel shield machine
隧道贯通	tunnel breakthrough
隧道开挖法	tunnel boring machine (TBM) method
隧道控制点	tunnel control station
隧道入口	adit of the tunnel
穗帽变换	tesseled cap transformation
缩放	zoom
缩放仪,缩图器,缩放	pantograph
缩微地图	microfilm map
缩微摄影	microcopying, microphotography
缩小	zoom in
索佳 GPS 接收机随机软件	GSSP, Spectrum Survey
所有权,拥有权	ownership
锁相环	phase lock loop(PLL)

T

塔	tower
塔顶	towertop
台链	station chain
太阳潮	solar tide
太阳辐射	solar irradiance, solar radiation
太阳辐射波谱	solar radiation spectrum
太阳光谱	solar spectrum
太阳罗经	sun compass
太阳能板	solar panel
太阳能建筑物	solar construction
太阳视差	solar parallax
太阳同步卫星	sun-synchronous satellite
太阳系	solar system
太阳系(力学)坐标系	Barycentric or dynamical coordinate system
太阴潮	lunar tide
太阴时	lunar time
泰勒级数	Taylor series
泰勒展开式	Taylor expansion
弹力	elastic force
弹性	elasticity
弹性变形	elastic deformation
弹性常数	elastic constant
弹性沉降	elastic settlement

T

弹性碰撞	elastic collision
弹性破坏	elastic failure
弹性体	elastic body
探测模式	detecting model
探测器	detector
探测信号	detected signal
探地雷达	ground penetrating radar
探空	radiosonde
探元	detector
探月	lunar exploration
探月航天器	lunar spatial probes
特解	particular solution
特征,标记,符号	signature
特征编码	characteristic coding,feature coding
特征点	feature points
特征方程	characteristic equation
特征分割	characteristic segmentation
特征根	characteristic root
特征函数	characteristic function
特征矩阵	characteristic matrix
特征聚类	feature cluster
特征空间	feature space
特征量积	characteristic product
特征码	feature code
特征码清单	feature code menu
特征匹配/基于特征的匹配	feature based matching

T

特征融合	character merger, feature fusion
特征水位	characteristic level of water
特征提取	feature extraction
特征选择	feature selection
特征组合	feature combination
特种地图	particular map
梯度,坡度,斜率	gradient
梯度向量	gradient vector
梯矩阵	echelon matrix
梯形	trapezium
梯形法则	trapezoidal rule
梯形集合	trapezoidal integration
梯阵式	echelon form
提取	extraction
提升算法	lifting scheme
提升小波	lifting scheme wavelet
体带调和函数,体带谐函数	solid zonal harmonics
体积计算	volume calculation
体视化	volume rendering
体素	voxel
天宝 GPS 随机数据处理软件	TGO, GPSurvey
天底,最低点	nadir
天底观测	nadir observation
天顶	zenith
天顶距	zenith distance, zenith angle
天极	celestial pole

天球	celestial sphere
天球赤道	celestial equator
天球纬圈	celestial parallel
天球子午线	celestial meridian
天球坐标	celestial coordinate
天球坐标系	celestial coordinate system
天然地形	natural terrain
天体	celestial body, celestial object
天体测量轨道	astrometric orbit
天体测量学	astrometry
天体力学	celestial mechanies
天体摄影仪	astrograph
天体图	celestial map
天文测定	astronomical determination
天文潮	astronomical tide
天文大地测量的	astrogeodetic
天文大地测量学	astronomical geodesy
天文大地垂线偏差	astro-geodetic deflection of the vertical
天文大地基准	astrogeodetic datum
天文大地水准测量	astro-geodetic leveling
天文大地网	astrogeodetic network
天文大地网平差	adjustment of astrogeodetic network, astrogeodetic adjustment
天文点	astronomical point
天文定位	astronomical positioning
天文定向	astronomical orientation
天文动力学,星际航行动力学	astrodynamics

天文方位角 astronomical azimuth
天文观测 astronomical observation
天文经度 astronomical longitude
天文经纬仪 astronomical theodolite
天文控制 astronomical control
天文量测 astronomical measurement
天文罗盘 astrocompass
天文年历 astronomical almanac, astronomical ephemeris
天文三角形 astronomical triangle
天文摄影机 astronomical camera
天文纬度 astronomical latitude
天文纬圈 astronomical parallel
天文位置 astronomical position
天文学 astronomy
天文重力水准测量 astro-gravimetric leveling
天文子午圈,天文子午线 astronomical meridian
天文坐标 astronomical coordinates
天线高度 antenna height
天线相位中心 antenna phase center
田球调和函数,田球谐函数 tesseral spherical harmonics
填充地图 outline map [for filling]
填充体系 infill system
填挖方量 cut-fill volume
填挖方平衡 cut and fill balance
条带测深仪 swath echo sounder

T

条带去除	destriping
条带噪声	streaking noise, striping noise
条件不等式	conditional inequality
条件方程	condition equation
条件分布	contional distribution
条件概率	conditional probability, contional probilitity
条件恒等式	conditional identity
条件模式谱	conditioned pattern spectrum
条件平差	condition adjustment
条件期望	contional expectation
铁路工程测量	railroad engineering survey
铁路曲线半径	radius of railway curve
停车场	car park
通道响应函数	band response function
通风孔	air drain
通风竖井	air shaft
通解,一般解	general solution
通信控制器	communication control unit
通用横轴墨卡托投影	Universal Transverse Mercator projection(UTM)
通用极球面投影	Universal Polar Stereographic projection(UPS)
同步测量实验	synchronous experiment
同步的,同时的	simultaneous, synchronous
同步观测	simultaneous observation
同步验潮	tide synobservation
同构	isomorphism

同名(像)点	homologous points
同名点的,同调的,类似的	homologous
同名光线,共轭射线	conjugate ray, corresponding image rays
同名核线	corresponding epipolar line
同名射线	homologous ray
同名像点	corresponding image points, homologous image points
同色异谱	metamerism
同态	homomorphic
同态滤波	homomorphic filter
同心圆	concentric circles
同心圆族	family of concentric circles
同轴电缆	coaxial cable
统计比值差值排序滤波器	SRROD filter
统计地图	statistical map
统计独立	statistical independence
统计分析	statistical analysis
统计检验	statistical testing
统计拟合拉伸	statistical fitting and stretching
统计数据	statistical data
统计显著性	statistical significance
统计学	statistics
投标	bidding
投射角	angle of projection
投射平面	projection plane

投射线	projection lines
投影(映)	projection
投影变换	projection transformation
投影变形	projection distortion
投影差	relief displacement
投影尺度,投影比例尺	projected scale
投影点密度	density of projected points
投影方程	projection equation
投影器	projector
投影仪	camera of projection
投影中心	aspect of projection
透明正片	diapositive
透射,透光率	transmittance
透射电子显微镜	transmission electron microscope
透射谱	transmission spectrum
透声窗	acoustic window
透视,透视图,透视的	perspective
透视截面法	perspective traces
透视投影	perspective projection
透视图	perspective view
透视中心	perspective centre
突然事件,紧急事件	emergency
图(形),数字	figure
图表	diagram
图层	coverage
图符,符号代码	code of symbols
图幅	mapsheet

T

图幅编号	sheet designation, sheet number
图幅接边	edge matching
图幅接合表	index diagram, sheet index
图幅中心	center of sheet
图根三角测量,低等三角测量	detailed triangulation
图号	sheet number
图号系统	sheet number system
图解	graphical solution
图解导线测量	graphical traversing
图解精度	graphical accuracy
图解纠正	graphical rectification
图解平差法	graphical adjustment
图廓	map border
图廓,边缘,界线	border
图廓分度	border division
图廓覆盖	map cover
图廓数据	block figure
图廓注记	border information
图例	lengend, map legend
图名	map title
图示,以图样表达	graphical representation
图像编码	image coding
图像变换	image transformation
图像插值	image interpolation
图像处理	image processing
图像传输	image transmission
图像分割	image segmentation

T

图像复合　image overlaying
图像几何纠正　geometric rectification of imagery
图像几何配准　geometric registration of imagery
图像降噪　image denoising
图像金字塔分解　image pyramid decomposition
图像空间分辨率　image spatial resolution
图像空间序列　sequence image segmentation
图像理解　image understanding
图像描述　image description
图像模拟　image simulation
图像配准　image registration
图像融合　image overlaying
图像识别　image recognition
图像数据　image data
图像数字化　image digitization
图像缩放　image resizing
图像条带　image strip
图形　graphics
图形(结构)问题　configuration problem
图形-背景辨别　figure-ground discrimination
图形叠置　graphic overlay
图形符号　graphic symbol
图形记号　graphic sign
图形数据　graphic data
图形数据库　graphic database
图形图像控制点　graph and image control point
图形纹理投影器　pattern texture projector
图形信息　graphic information

图形要素　pattern element
图形用户接口　graphic user interface(GUI)
图形元素　graphic element
图载水深　charted depth
图质改善处理　image quality correction
土坝,土堤　earth dam
土堤　earth embankment
土地保护　land conservation
土地测量　land survey
土地出让　land leasing
土地调查　land investigation
土地分割　land division
土地分类　land classification
土地覆盖　land cover
土地估价,土地评估　land appraisal,land evaluation
土地管理　land management
土地规划测量　land planning survey
土地开发　development of land
土地利用调查　land use survey
土地利用动态监测　land use monitoring
土地利用分析　land use analysis
土地利用规划　land use planning
土地利用现状图　present landuse map
土地平整　land-leveling operation
土地使用权　right of land usage
土地所有权　land ownership
土地退化　land degradation
土地信息系统　land information system(LIS)

T

土地征用	land acquisition
土地转让	land transfer
土地资源	land resources
土方计算	cut and fill estimate
土方量	earthwork volumes
土木工程	civil construction
土壤含水量	soil water content
土壤含盐量	soil salt
土壤侵蚀强度	soil erosion intensity
土壤热惯量	soil thermal inertia
土壤水分	soil moisture
土壤温度	soil temperature
推荐航线	recommended route
推理判决	inference decision
推论	deduction, inference
推扫式传感器	push-broom sensor
推扫式遥感影像	satellite imagery of push broom
推算船位	estimated position
推算航程	estimated distance
推算航向	estimated course
退卷积	deconvolution
托帕克斯卫星	TOPEX/POSEIDON(T/P)
拖底扫海	drag sweep, ground sweeping
拖索	drag wire
拖尾	tails off
拖曳式换能器	tow transducer
拖曳载体	tow vehicle
陀螺	gyro

T

陀螺摆动	gyro oscillation
陀螺北	gyrocompass north
陀螺方位角	gyro azimuth
陀螺分画板	gyro scale
陀螺光标线	gyro mark
陀螺经纬仪	gyro theodolite, gyroscopic theodolite
陀螺罗航向	gyro course
陀螺罗经	gyrocompass
陀螺稳定平台	gyro-stabilized platform
陀螺旋转轴	gyro spin axis, rotating axis of the gyro
陀螺仪,回转仪	gyroscope
陀螺仪定向	gyroscopic orientation
陀螺仪定向测量	gyroscopic orientation survey
椭球,椭球体	ellipsoid
椭球扁率	flattening of ellipsoid
椭球长半轴,地球长半轴	semi-major axis of ellipsoid
椭球调和函数,椭球谐函数	ellipsoidal harmonics
椭球短半轴,地球短半轴	semi-minor axis of ellipsoid
椭球法线	ellipsoidal normal
椭球面方位角	spherical azimuth
椭球面曲率	ellipsoidal curvature
椭球偏心率	eccentricity of ellipsoid
椭球体面距离	ellipsoidal distance

椭球体面坐标	ellipsoidal coordinates
椭球子午线	ellipsoidal meridian
椭圆	ellipse
椭圆定位	elliptic positioning
椭圆拟合	best fitting ellipse
拓扑地图	topological map
拓扑分析	topological analysis
拓扑关系	topological relation
拓扑检索	topological retrieval
拓扑结构	topological structure
拓扑数据模型	topological data model
拓扑学,地志学	topology

T

W

挖掘工程	excavation works
挖土机	excavator
外表裂缝	external crack
外部定向	exterior orientation
外部精度	external accuracy
外部可靠性	external reliability
外部支撑	external bracing
外定标	vicarious external calibration
外定向	exterior orientation
外方位元素	exterior orientation element, exterior orientation parameters
外核	outer core
外加荷载	applied load
外加荷载,附加荷载	imposed load
外角	exterior angle
外接圆	circumcircle, circumscribed circle
外接圆半径	circumradius
外接圆心	circumcentre
外力	external force
外矢距	external distance
外业	fieldwork
外业步骤	field procedures
外业草图	field sketch
外业规范,外业手册	field manual

W

外业检核	field check
外业手簿	field book
外业数据	field data
外直径	external diameter
完好性	integrity
完全规范化调和函数,完全规范化谐函数	fully normalized harmonics
完全规范化球谐函数	fully normalized spherical harmonics
完全运行能力	fully operational capability(FOC)
万维网地理信息系统	web GIS
万有引力定律	law of universal gravitation
网,网络	network
网格	grid
网格单元	grid cell
网格地图	grid map
网格法	grid method
网格计算	grid computing
网格技术	grid technology
网格结构	grid structure
网络 RTK	network RTK
网络安全	safety of network
网络分析	network analysis
网络联结	network link
网平差	network adjustment
网纹片	transparent foil
网线	ruling

W

网状结构	net structure
往返测水准测量	double leveling, double-run leveling, reciprocal leveling
往复流	reversing current
望远镜	telescope
危险界限	limiting danger line
危险水深	danger sounding
微波测距仪	microwave distance measuring instrument, microwave rangefinder
微波辐射	microwave radiation
微波辐射计	microwave radiometer
微波极化比	microwave polarization ratio
微波扫描仪	microwave scanner
微波图像	microwave imagery
微波遥感	microwave remote sensing
微波遥感器	microwave remote sensor
微波遥感卫星,1995年发射	RadarSat 2
微波遥感卫星,2000年发射	SRTM(The Shuttle Radar Topography Mission)
微波遥感卫星,2001年发射	ENVISAT
微波遥感卫星,2003年发射	ALOS
微差水准测量	differential leveling
微调,细调;精密平差	fine adjustment
微分	differential
微分法	differentiation

微分方程　differential equation
微分纠正　differential rectification
微分中值定理　differential mean value theorem
微观照片　microscopic photograph
微积分学　calculus
微重力测量　microgravimetry
韦史巴赫三角形　Weisbach triangle
韦斯四边形法　weiss quadrilateral
围堰　cofferdam
唯一解　unique solution
唯一性　uniqueness
维(数)　dimension
维纳频谱　Winer spectrum
维宁曼尼斯公式　Vening-Meinesz formula
伪彩色　pseudo-color
伪彩色图像　pseudo-color image
伪角点　false corner
伪距　pseudo-range
伪距测量　pseudo-range measurement
伪随机噪声　pseudo-random noise (PRN)
伪随机噪声码　pseudo-random noise code
伪卫星　pseudolite
尾次回波　last echo(return)
纬差　difference of latitude
纬度　latitude
纬度误差　error in latitude
纬圈,平行圈　latitude circle, parallel circle
卫星测高　satellite altimetry

卫星成像	satellite images
卫星城	satellite town
卫星大地测量,卫星大地测量学	satellite geodesy
卫星定位	satellite positioning
卫星定向	satellite orientation
卫星多普勒[频移]测量	satellite Doppler shift measurement
卫星多普勒定位	satellite Doppler positioning
卫星高度	satellite altitude
卫星高度计	satellite altimeter
卫星跟踪卫星技术	satellite-to-satellite tracking(SST)
卫星跟踪站	satellite tracking station
卫星共振分析	analysis of satellite resonances
卫星构形	satellite configuration
卫星轨道	satellite orbit
卫星轨道改进	improvement of satellite orbit
卫星海洋学	satellite oceanography
卫星激光测距	satellite laser ranging(SLR)
卫星激光测距仪	satellite laser ranger
卫星扫描带,条带	swath
卫星摄影	satellite photography
卫星摄影测量	satellite photogrammetry
卫星受摄运动	perturbed motion of satellite
卫星星下点	sub-satellite point
卫星星座	satellite constellation
卫星遥感测深	satellite remote sensing fathoming
卫星影像	satellite image

卫星运动方程	equation of motion of the satellite
卫星钟	satellite clock
卫星重力梯度测量	satellite gradiometry
卫星姿态	satellite attitude
未训练类别	untrained types
未知点位	unknown position
未知量(参数)	unknow quantity(parameter)
未知数	unknown
位移	displacement
位移传感器	displacement transducer
位移观测	displacement observation
位置	position
位置(线交)角	intersection angle of LOP
位置测定	determination of position
位置传感探测器	position sensitive detector(PSD)
位置格式	position format
位置函数,坐标函数	position function
位置和姿态	position and attitude
位置精度	positional accuracy
位置精度因子	position dilution of precision (PDOP)
位置面	surface of position(SOP)
位置平差	adjustment of position
位置平均误差	average error in position
位置图	location map
位置线	line of position (LOP)
位置线方程	equation of LOP
位置向量	position vector

W

温差 temperature difference
温差改正 temperature correction
温度传感器 temperature sensor
温度计 thermograph
温度廓线 temperature profile
温度跃层 thermocline
温-盐图解 temperature/salinity diagram
纹理 texture
纹理分割 texture image segmentation
纹理分类 texture classification
纹理分析 texture analysis
纹理谱 texture spectrum
纹理特征 texture feature
纹理图像复原 texture restoration
纹理压缩 texture compression
纹理影像 texture image
纹理映射 texture mapping
纹理增强 texture enhancement
纹理指数 texture index
稳定点 stable point
稳定锚固点 stable anchor point
稳定平台 stabilized platform
稳定性 stability
稳健统计学 robust statistics
稳健性 robustness
稳态海面地形 stationary sea surface topography
沃尔什变换 Walsh transformation
污水系统 sewage system

无抽样小波变换	undecimated wavelet transform
无电离层组合	ionosphere-free combination
无定向重力仪	astatic gravimeter
无反射全站仪	reflectorless total station
无缝拼接	seamless contiguity
无缝数据库	seamless database
无缝镶嵌	seamless registration
无关观测值	uncorrelated observations
无界函数	unbounded function
无控制点镶嵌图,像片略图	uncontrolled mosaic
无理方程	irrational equation
无理数	irrational number
无扭	torque eliminated
无偏估计	unbiased estimate
无偏估计量	unbiased estimator
无人机遥感监测系统	unmanned air vehicle for remote sensing system
无人机遥感系统	UAVRSS(unmanned aerial vehicle remote sensing system)
无人驾驶飞行器	UAV(Unmanned Aerial Vehicle)
无损压缩	lossless compression
无线电波	radio wave
无线电测高仪	radio altimeter
无线电大地测量,雷达三角测量	radio triangulation
无线电定位	radio positioning
无线电航行警告	radio navigational warning

W

无线电指向标	radio beacon
无线电指向标表	list of radio beacon
无线网地理信息系统	wireless aplication protocol GIS (WAP GIS)
五角棱镜	pentaprism
物方影像匹配	object space image matching
物镜分辨率	resolving power of lens
物空间坐标系	object space coordinate system
物理大地测量学	physical geodesy
物理海洋学	physical oceanography
物理学	physics
物资储备	material storage
误差	error
误差传播	error propagation
误差方程	error equation
误差分布	error distribution
误差分析	error analysis
误差估计	error assessment
误差估算	error estimation
误差检验	error test
误差理论	theory of errors
误差椭圆	error ellipse
误差项	error term

X

吸声器	absorber
吸收	absorption
吸引力	attraction, attractive force
希尔伯特-黄变换	Hilbert-Huang transform
稀疏化处理	sparseness
系列地图	series maps
系数	coefficient
系统采样	systematic sampling
系统集成	system integration
系统控制中心	system control center(SCC)
系统误差	systematic error
细化算法	thinning algorithm
峡谷形态	gully shape
狭义空间信息网格	specialized information grid
狭义相对论	special relativity
下标	suffix
下部结构,基础,地下建筑	substructure
下垂,下陷,垂度	sag
下三角阵	lower triangular matrix
下限	lower limit
下中天	lower transit
先验	priori
先验方差因子	priori variance factor

先验概率	priori probability
先验知识	priori knowledge
闲置地,空地	vacant land
弦长	chord length
弦线支距法	chord offset method
显函数	explicit function
显微摄影	photomicrography
显微摄影测量	microphotogrammetry
显著性检验	significance test, test of significance
显著性水平	significance level
险恶地	foul ground
现场安装	on-site installation
现场边界(线)	site boundary
现场计算	field calculation
现场勘测	site reconnaissance
现场施工	in-place construction
现代风格	up-to-date style
现代文明	modern civilization
现浇混凝土	cast-in-place concrete
现势地图	up-to-date map
现势资料	current information
现有建筑	existing construction
线段	line segment
线段匹配	line match
线划地图	line map
线矩	line moment
线扩散函数	line diffusion function model

线列水听器 line hydrophone
线路测量 route survey
线路测设 route stationing
线路工程测量 route engineering survey
线路平面图 route plan
线路水准测量 route leveling
线路中线测量 location of route
线扫描 line scanning
线图层 line coverage
线纹米尺,日内瓦尺 standard meter
线形锁 linear triangulation chain
线形网 linear triangulation network
线性调频信号 linear frequency modulation signal
线性多尺度变换 linear multiscale transform
线性方程 linear equation
线性非平稳过程 linear nonstationary process
线性光谱混合求解方法 linear spectral unmixing
线性过程 linear process
线性化的数学模型 linearized math model
线性回归 linear regression
线性滤波模型 linear filter model
线性滤波器 linear filter
线性模型 linear models
线性平稳模型 linear stationary model
线性收敛性 linear convergence
线性微分方程 linear differential equation
线性无关的 linearly independent

线性无偏估计	linear unbiased estimator
线性相关	linear correlation
线性相关的	linearly dependent
线阵 CCD 推扫式	linear CCD push-broom
线阵列	linear array
线阵推扫式影像	linear array push-broom imagery, linear pushbroom imagery
线阵遥感器	linear array sensor, push-broom sensor
线中心投影	line perspective
线状地物	line-like object
线状符号	line symbol
线状目标	line target
限差,容许误差	tolerance
限差范围	tolerance radius
限航区	restricted area
乡村规划	rural planning
相对(点位)误差椭圆	relative (point) error ellipse
相对定位	relative positioning
相对定向	relative orientation
相对定向元素	element of relative orientation
相对辐射校正	relative radiometric correction
相对高度	relative altitude
相对极大	relative maximum
相对极小	relative minimum
相对精度	relative accuracy
相对控制	relative control
相对论改正	relativistic correction

X

相对论效应　relativistic effect
相对频数　relative frequency
相对倾角　relative tilt
相对视差　relative parallax
相对速度　relative velocity
相对位移　relative displacement
相对位置　relative position
相对误差　relative error
相对运动　relative motion
相对重力　relative gravity
相对重力测量　relative gravity measurement
相对重力仪　relative gravimeter
相干度　coherence level
相干散射模型　coherent scattering model
相干条纹　interferogram fringe
相干系数　coherence coefficient
相关　correlation
相关观测值　correlated observations
相关配准　correlation registration
相关平差　adjustment of correlated observations
相关器　correlator
相关松弛法　correlation relaxation
相关系数　correlation coefficient
相关性,相对论　relativity
相机标定,摄像机标定　camera calibration
相邻航线　adjacent flight line

相似三角形 similar triangles
相似图形 similar figures
相似性 similarity
相似性度量 similarity measure
相速 phase velocity
相位,相 phase
相位补偿器 phase compensator
相位传递函数 phase transfer function(PTF)
相位解缠 phase unwrapping
相位模糊度 phase ambiguity
相位模糊度解算 phase ambiguity resolution
相位漂移 phase drift
相位平滑伪距 phase-smoothed pseudo-range
相位稳定性 phase stability
相位周 phase cycle
相位周值 phase cycle value
相移 phase shift
相异根 distinct roots
相异解 distinct solution
镶嵌 mosaic
镶嵌处理 mosaic assembling
镶嵌索引图 index mosaic
详图,细部图 detail drawing
详细规划 detailed planning
详细设计 detail design
向量 vector
向斜 syncline
向心加速度 centripedal acceleration

X

向心力	centripetal force
项目工程师	project engineer
项目管理	project management
项目规划	project planning
项目监督	project supervision
项目设计	project design
象限	quadrant
象形符号	replicative symbol
象形图	pictogram
像差改正	correction for aberration
像场角	angular field of view, objective angle of image field
像等角点	isocenter of photograph
像底点	photo nadir point
像地平线,合线	image horizon, horizon trace, vanishing line
像点	image point
像点横坐标	abscissa of image point
像点位移	displacement of image point
像幅	picture format
像空间	image space
像空间坐标系	image space coordinate system
像片	photo, photograph
像片比例尺	photo scale
像片地质判读,像片地质解译	geological interpretation of photograph
像片方位角	azimuth of photograph
像片方位元素	photo orientation elements

像片纠正	photo rectification
像片内方位元素	elements of interior orientation
像片判读	photo interpretation
像片平面图	photoplan
像片倾角	tilt angle of photograph
像片倾斜	photo tilt
像片外方位元素	elements of exterior orientation
像片镶嵌	photo mosaic
像片旋角	swing angle, yaw
像片主距	principal distance of photo
像片坐标,像点坐标	image coordinate
像平面	image plane
像平面坐标系	photo coordinate system
像素	picture element/ pixel
像素定位	pixel location
像移	image shift
像移补偿	image motion compensation(IMC)
像元,像素	pixel
像元分解	pixel unmixing
像元光谱混合	spectral mixing
像元畸变校正	distorted pixel correction
像质评价	image quality measure
像主点	principal point of photograph
像主纵线	principal line [of photograph]
消法	elimination
消防栓,消防龙头	hydrant
消声	noise elimination
消元法	elimination method

小比例尺 small scale
小波包 wavelet packet
小波包变换 wavelet-packet transform
小波变换 wavelet transform
小波变换法 wavelet transform method
小波插值 wavelet interpolation
小波分析 wavelet analysis
小波系数标准方差 wavelet coefficient standard deviation
小波直方图 wavelet histogram
小潮 neap tide
小潮平均低潮位 mean low water neaps
小潮平均高潮位 mean high water neaps
小角度法 angle method, minor angle method
小角度倾斜 low obligue
小面阵 CCD 影像 small matrix array image
小数 decimal
小数点 decimal point
小像幅航空摄影 small format aerial photography (SFAP)
小样本 small-sample
校园规划 campus planning
校园平面图 campus plan
协方差函数 covariance function
协议地球参考系 conventional terrestrial reference system (CTRS)
协议惯性参考系 conventional inertial reference system (CIRS)

协因数矩阵	cofactor matrix
斜边	hypotenuse
斜高	slant height
斜井	inclined shaft
斜距	slope distance
斜拉桥	cable-stayed
斜面	inclined plane
斜轴投影	oblique projection
泄水建筑物	discharge structure
心理救助	psychological rescue
心象地图	mental map
新版海图	new edition chart
新建工程	new construction
新区	newly-built district
信标	beacon
信号分频器	antenna splitter
信号源,辐射体	source
信号障碍	signal obstruction
信息安全	information safty
信息复合	information compound
信息干扰事件	information disturbance accident
信息共享	information sharing
信息获取	information acquisition
信息融合	information fusion
信息熵	information entropy
信息时代	information age
信息提取	information extraction
信息网格	information grid

信息隐藏	information hiding
信息属性	information attribute
信源编码	source coding
信噪比	signal-to-noise ratio
星基增强系统	satellite-based augmentation system (SBAS)
星空夜视仪	astrolight viewer
星历	ephemeris
星历误差	ephemeris error
星载	space-borne
星载合成孔径雷达	spaceborne SAR
星载加速度传感器	satellite acceleration sensor
星载微波辐射计	space microwave radiometer
星载遥感器	satellite sensor
星座	constellation
行程编码,游程编码	run length encoding
行程图	travel graph
行抖动差	line jitter
行间传输	interline transfer
行列式	determinant
行树与篱笆	tree rows and hedges
行为反应	behavior reaction
行向量	row vector
行星大地测量学	planetary geodesy
行星岁差	plantary precession
行星天文学	plantary astronomy
行星学	planetology
行政区划图	administrative map

X

形式化描述	formalization description
形态变换	morphology transformation
形态梯度	morphological gradient
形态小波变换	morphological wavelet
形态学	morphology
形状	shape
形状特征	shape features
修版	retouching
修正	amend
虚部	imaginary part
虚地图	virtual map
虚根	imaginary root
虚拟参考站	virtual reference station(VRS)
虚拟城市	virtual city
虚拟地景	virtual landscape
虚拟现实	virtual reality(VR)
虚拟现实技术	virtual reality technology
虚拟真二维控制场	virtual real-2D field
虚数	imaginary number
序贯法	sequential approach
序贯平差	sequential adjustment
序列影像	image sequence
蓄洪区	flood storage works
悬臂	cantilever
悬臂梁	cantilever beam, overhanging beam
悬带	suspension tape(wire)
悬点	point of suspension
悬浮泥沙	suspended sediment concentration

悬挂式陀螺	suspended gyro
悬链	catenary
悬索桥	suspension bridge
旋角校正	swing adjustment
旋镜,旋转镜,回转反射镜	rotating mirror
旋转	revolution, rotation
旋转不变	rotation invariant
旋转参数	rotation parameters
旋转方向	direction of spin
旋转角	angle of rotation
旋转角动量	spin angular momentum
旋转矩阵	rotation matrix
旋转速度	spin speed
旋转体	solid of revolution
旋转陀螺	spinning gyro
旋转椭球	spheroid of revolution
旋转椭球体	ellipsoid of revolution
旋转中心	centre of rotation
旋转轴	spin axis
旋转轴,公转轴	rotation axis
选权迭代法	iteration method with variable weights
选线	route selection
选择可用性	selective availability (SA)
选址	site selection
学习算法	learning algorithm
雪盖	snowpack

雪盖提取	snow-cover extracting
寻北	north-seeking
寻北仪器	north-finding instrument, polar finder
循环小数	recurring decimal
训练	training
训练区采样	sampling for training areas
训练样本	training samples

Y

压电换能器	piezoelectric transducer
压力(压强)计	piezometer
压力式自计验潮仪	pressure autorecording tide gauge
压力验潮仪	pressure gauge
压缩估计	stein estimator
压缩域	compressed domain
亚均值滤波	sub-mean filter algorithm
亚像元	sub_pixel
亚洲沙尘暴	Asian dust
延迟锁相环	delay lock loop(DLL)
严格单调	strictly monotonic
严格单调函数	strictly monotonic function
严密平差	rigorous adjustment
岩石圈结构	lithospheric structure
岩土工程	geotechnical
岩土工程方法	geotechnical method
岩性分类	rock classification
沿岸测量,沿海测量	coastal survey,coastwise survey
沿岸航行图	coastwise navigation
沿岸流	coastal current
沿计划航线上的航速	velocity made good (VMG)
沿径分量	radial component
盐度计	salinometer
盐度跃层	halocline

盐度准确度	accuracy of sea water salinity
颜色矩	color moments
颜色空间	color space
颜色空间转换	color space transformation
颜色扭曲	color distortion
掩膜技术	masking technique
眼基线调节	interpupilliary adjustment
演绎	deduce
演绎推理	deductive reasoning
验潮	tide gauge
验潮井	tidal well
验潮水尺	tidal staff
验潮站	tidal station
验潮站水位标志	tidal station level mark
验算	checking
洋流	ocean current
样本空间	sample space
样本预选取	pre-selection sample
样区	ROI
样条函数法	spline method
遥测传感器,遥感器	remote sensor
遥感	remote sensing(RS)
遥感测深	remote sensing sounding
遥感反演	remote sensing retrieval
遥感监测	remote sensing monitoring
遥感勘查	remote sensing exploration
遥感模式识别	pattern recognition of remote sensing

遥感判读	remote sensing interpretation
遥感平台	remote sensing platform
遥感数据获取	remote sensing data acquisition
遥感图像模拟	simulation of RS image
遥感信源	remote sensing information source
遥感制图	remote sensing mapping
遥控器	remote controller
要素,特征	feature
要素标识码	feature identifier
要素分类	feature catalogue
要素关系	feature relationship
要素类型	feature type
要素属性	feature attribute
野外地质图	geological map
野外勘测	field reconnaissance
野外填图,制图	field mapping
液核	fluid outer core
液晶显示器	liquid crystal display (LCD)
液晶眼镜	crystal eyeglasses
液晶遮光眼镜	crystal shaded eyeglasses
液体静力水准测量	hydrostatic leveling
一般式,通式	general form
一等大地网	first-order geodetic network
一等导线	first-order traverse
一等精度	first-order accuracy
一等三角测量	first-order triangulation
一等三角网	primary triangulation net
一等水准	first-order leveling

一等水准点	first-order benchmark
一等水准网	first-order leveling network
一井定向	two-wire method: one shaft
一类设计:图形结构设计	first-order design (FOD): configuration problem
一致(的),均匀(的)	uniform
仪表盘,面板	panel
仪器常数	constant of instrument, instrumental constant
仪器对中	centering of instrument
仪器高	height of instrument (HI), instrument height
仪器检测	instrument testing
仪器精度	accuracy of instruments
仪器误差	instrumental error
仪器校正	instrument adjustment
仪器置平	leveling of instrument
移动拟合法	moving average method
移动曲面拟合法	fitting of moving quadric surface
移位视差/伪视差	motion parallax / pseudo parallax
移项,转置	transpose
遗传算法	genetic algorithm(GA)
已知站(点)	known station
蚁群算法	ant algorithm
蚁群行为仿真	simulation of ant colony behavior
异常位,扰动位	anomalous potential, disturbing potential
异常值,粗差	blunder, gross error, outlier

Y

异轨遥感影像	remote sensing images from different orbits
异质性	heterogeneity
译码,解码	decoding
溢流装置	excess flow device
铟瓦尺	invar rod
铟瓦基线尺	invar baseline wire
铟瓦条形码尺	invar barcode rod
阴极射线	cathode-ray
阴像	negative image
阴影补偿	shadow compensation
阴影分析	shadow analysis
阴影检测	shadow detection
阴影去除	shadow removal
阴影提取	shadow extraction
阴影校正	shadow correction
阴影增强	shadow enhancement
银河系,星系	galaxy
引潮力	tide-generating force
引潮位	tide-generating potential
引航图	pilot chart
引航图集	pilot atlas
引力	gravitation
引力常数,重力常数	gravitational constant
引力场	gravitational field
引力加速度	gravitational acceleration
引力球谐函数	gravitational harmonics
引力势,引力位	attraction potential

Y

引力位 gravitational potential
引力位函数 gravitational potential function
引力向量,重力向量 gravitational vector
引桥 approach
引水锚地 pilot anchorage
引张线法 method of tension wire alignment
隐函数 implicit function
隐患 hidden trouble
印刷版 printing plate
印刷海流计 chart current meter
应变计,张力计 strain gauge
应变片花 strain rosette
应变椭圆 strain ellipse
应变张量 strain tensor
应急 emergency response
应急处置 emergency disposal
应急管理 emergency management
应急演习 emergency exercise
应急预案,应急计划 emergency plan
应急准备 emergency preparedness
应用程序接口 application programming interface (API)
荧光地图 fluorescent map
荧光遥感 fluorescence remote sensing
影像,图像 image,imagery
影像插值 image interpolation
影像地图 photomap
影像地质图 geological photomap

影像分辨率 image resolution
影像分块 image blocking
影像分类 image classification
影像分析,图像分析 image analysis
影像复合 image integration
影像复原 image restoration
影像几何纠正 geometric correction of image
影像金字塔 image pyramid
影像精化 image refinement
影像纠正 image rectification
影像块编码 image block coding
影像判读 image interpretation
影像匹配 image match
影像清晰度 image acutance
影像融合 image fusion
影像融合评价 image fusion evaluation
影像色彩失真校正 color distortion correction of image
影像数据库 image database
影像图 image map
影像位移,像移 image displacement
影像纹理分类 classification of image texture
影像细部,图像细部 photographic detail
影像相关 image correlation
影像镶嵌 image mosaic
影像信息认知 image information cognition
影像压缩 image compression
影像增强 image enhancement
影像遮蔽 occlusion

影像质量	image quality
影像质量控制	image quality control
影像重建	image reconstruction
硬拷贝	hard copy
硬式扫海具	bar sweeper
永久测站,埋石测站,埋石点	permanent station
永久建筑	permanent building
永久性测量标	permanent geodetic marker
永久性结构(建筑物)	permanent structure
涌浪滤波补偿器	roller filtering compensator
用户部分	user segment
用户等效距离误差	user equivalent range error
用户精度	user accuracy
用水量	water supply volume
优化设计方法	approaches to the optimal design
游标精度	vernier accuracy
游艇用图	smallcraft chart,yacht chart
有毒化学品灾害	venomous chemical accident
有毒气体	dangerous gases
有界函数	bounded function
有界序列	bounded sequence
有控制点镶嵌图,控制镶嵌	controlled mosaic
有理函数	rational function
有理化	rationalization
有理数	rational number
有损压缩	lossy compression

有限概率空间	finite probability space
有限级数	finite series
有限集	finite set
有限维向量空间	finite dimensional vector space
有限元法	finite element method
有效半径	effective radius
有效发射率	effective emissivity
有效范围	effective range
有效估计量	efficient estimator
有效期限	term of validity
有效数字	significant figure
有效性	validity
右边	right femur
右手边	right-hand-side(R. H. S)
右手坐标系	right-handed coordinate systems
右转角	angle-to-right
余割	cosecant
余函数	complementary function
余角	complementary angle
余切	cotangent
余弦	cosine
余弦法 / 矢量法(基于角锥体原理的方法)	cosine method / vector method
余弦公式	cosine formula
余弦散射模型	cosine backscatter model
鱼礁	fishing rock
鱼型浮	fish buoy

Y

鱼堰	fish haven
鱼栅	fish stacks
渔业用图	fishing chart
宇宙飞船	spacecraft
宇宙制图	cosmic mapping
语义共享	semantics sharing
语义提取	semantic extraction
语义信息	semantic information
预报地图	prognostic map
预处理	pre-processing
预打样图	pre-press proof
预警系统	pre-warning system, risk warning system
预览	preview
预期精度	expected accuracy
预期值	expected value
预设标志	premarking
预应力混凝土	prestressed concrete, re-enforced concrete, reinforced concrete
预制符号	preprinted symbol
预制感光板,PS 版	presensitized plate
阈值	threshold / thresholding value
元数据	metadata
元数据模式	metadata schema
元数据实体	metadata entity
元数据元素	metadata element
元数据子集	metadata section
元素	element

原点	origin
原始数据	raw data
原始条件,初始条件	initial condition
原武汉测绘科技大学 GPS 数据处理软件	LIP
原子时	atomic time
原子钟	atomic clock
圆概率误差	circular error probable (CEP)
圆函数,三角函数	circular function
圆曲线测设	circular curve staking, setting out of circular curves
圆曲线几何形状	circular curve geometry
圆曲线要素	basic elements of a circular curve
圆水准器	circular bubble
圆-圆定位,距离-距离定位	range-range positioning
圆周	circumference
圆周运动	circular motion
圆柱调和函数,贝塞尔函数	cylindrical harmonics
圆柱体	cylinder
圆柱投影	cylindrical projection
圆柱形的	cylindrical
圆锥投影	conic projection
圆族	family of circles
远程电子测距仪	long-range EDM instrument
远程定位系统	long-range positioning system
远程教育	distance education

Y

远地点	apogee
远海测量	pelagic survey
远海航行图	pelagic sailing chart
远日点	aphelion
远洋航行图	oceanic sailing chart
约束	constraint
约束条件	constraint condition
约束最优化	constrained optimization
月面测量,月面测量学	selenodesy
月面测量学	lunar geodesy, selenodesy
月平均海面	monthly mean sea level
月球地形测绘	lunar topography surveying
月球轨道飞行器	lunar orbiter
月球视差	lunar parallax
月球投影经度	lunar longitude
月球纬度	lunar latitude
月球中天	moon's transit
跃变层深度图	layer depth chart
云层检测	cloud detection
云覆盖	cloud-cover
云干扰	cloud contamination
云雪检测	cloud and snow detection
匀光	dodging
匀加速度	uniform acceleration
匀速度	uniform velocity
匀速率	uniform speed
匀速运动	uniform motion
匀质球体	homogeneous sphere

运动补偿	motion compensation
运动参数估计	shift estimation
运动对象	moving object
运动检测	motion detection
运动模型估计	motion estimate
运动目标检测	moving target detection
运输隧道	conveyance tunnel
运行成本,使用费	operating cost
运营和维护	operation and maintenance
晕滃法	hachuring
晕渲	brush shade
晕渲法	hill shading

Z

杂波	clutter
灾害管理	disaster management
载波	carrier
载波辅助跟踪	carrier-aided tracking
载波相位	carrier phase
载波相位测量	carrier phase measurement
再分结构	subdivisional organization
再生能源	renewable energe resource
在航	On-The-Fly（OTF）
在建项目	item under construction
在线访问	online access
暂住人口	temporary population
噪声	noise
噪声测量水听器	noise-measurement hydrophone
噪声定位	locating of noise pixel
噪声降低	noise reduction
噪声抑制	noise suppressing
增广矩阵	augmented matrix
增量	increment
增强	enchance
栅格化	rasterization
栅格绘图	raster plotting
栅格结构	raster structure
栅格-矢量转换	raster-to-vector conversion

栅格数据	raster data
窄带多光谱	narrow band multispectral
窄束回声测深仪	narrow beam echo sounder
窄相关器	narrow correlator
窄巷	narrow lane
展开式	expanded form
战场环境仿真	battlefield environment simulation
站心球面坐标	topocentric spherical coordinate
站心原点	topocentric origin
站心直角坐标	topocentric cartesian coordinate
站心坐标	topocentric coordinate
站心坐标系	topocentric coordinate system
张力	tension, tensional stress
张力,应变	strain
张量	tensor
章动	nutation
涨潮	flood tide
涨潮流	flood current
障碍物,栅栏	barricade
招标	bid call
照明工程	illuminanting engineering
照相制版镜头	printer lens
照准	collimation
照准部水准器,长水准器	long level bubble
照准点	tatget point
照准点归心	target centring
照准精度,目标指示精度	pointing accuracy

Z

照准轴 sighting axis
遮蔽补偿 occlusion compensation
遮挡关系 hiding relation
折光差改正 refraction correction
折射系数 coefficient of refraction
真北 true north
真彩色 true color
真春分点 true vernal equinox
真地平线,真水平线 true horizon
真方位角 true azimuth
真航向 true course
真恒星时 true sidereal time
真实孔径雷达 real-aperture radar
真矢量空间 real vector space
真太阳日 solar day, true solar day
真太阳时 solar time, true solar time
真误差 true error
真值 true value
真子午线 true meridian
阵列摄影机 array camera
阵列优化 array optimization
阵速水听器 velocity hydrophone
振动 oscillation
振动频谱 vibration spectrum
振动收敛性 oscillatory convergence
振幅离差指数 amplitude dispersion index
震源 seismic focus, hypocenter
震源发生器 seismic site producer

整平仪器	leveling the instrument
整数	integer
整数部分	integer part
整数解	integer solution
整数小波变换	integer wavelet transform
整数值	integer value
整体大地测量	integrated geodesy
整体构造活动	global tectonic activity
整体几何约束	global geometrical constraint
整体平差	simultaneous adjustment
整体松弛	global relaxation
整周模糊度解算	integer ambiguity resolution
正常尺寸,标准尺寸	normal size
正常动力高	normal-dynamic height
正常高	normal height
正常水椭球,水准椭球	normal level ellipsoid
正常位	normal potential
正常引力位	normal gravitational potential
正常正高	normal-orthometric height
正常重力	normal gravity
正常重力场	normal gravity field
正常重力公式	normal gravity formula
正常重力位	normal gravity potential
正常重力线	normal gravity plumb line
正锤[线]观测	direct plummet observation
正方形分幅	square map subdivision
正高	orthometric height

正规化	normalize
正交	orthogonal
正交变换	orthogonal transformation
正交小波变换	orthogonal wavelet transform
正交性	orthogonality
正立体效应	orthostereoscopy
正片,正	positive
正色片	orthochromatic film
正射纠正	ortho rectification
正射摄影	orthophotograph
正射投影	orthographic projection, orthophotography
正射透视,正射透视图	orthographic perspective
正射像片	orthophoto
正射像片镶嵌图	orthophotomosaic
正射影像地图	orthophoto map
正射影像	orthoimage
正射影像技术	orthophoto technique
正数	positive number
正态分布,常态分布	normal distribution
正态分布曲线	normal curve
正态随机变量	normal random variable
正态误差分布曲线	normal error distribution curve
正弦	sine
正弦公式	sine formula
正像,正片	positive image
正整数	positive integer

Z

正直摄影测量	vertical photogrammetry
正指数	positive index
正轴投影	normal projection
帧传输	frame transfer
帧-行间传输	frame interline transfer(FIT)
支撑体系	support system
支撑向量回归	support vector regression
支撑向量机,支持向量机	support vector machine(SVM)
支导线	open traverse
支点	fulcrum,pivot
支架	bracket
支水准路线	spur leveling line
支柱,压杆	pillar,strut
知识表示	knowledge representation
知识推理	knowledge reasoning
知识约简	attributes reduction
直方图	histogram
直方图不变矩	invariable moment of histogram
直方图规格化	histogram specification
直方图均衡	histogram equalization
直方图匹配	histogram matching
直角	right angle
直角坐标	rectangular coordinate
直角坐标网	rectangular grid
直角坐标系统	rectangular coordinate system
直接(解析)法	direct (mathematical) solution
直接对地目标定位	DG direct georeferencing

Z

直接法纠正	direct scheme of digital rectification
直接平差	direct adjustment
直接线性变换	direct linear transformation(DLT)
直立圆柱	right circular cylinder
直立圆锥	right circular cone
直射辐射	direct radiance
直伸导线	straight-line traverse
直线	straight line
直线抽取	straight-line extraction
直线导航	straight line navigation
直线检测	straight line detection
直线摄影测量	line photogrammetry
直线相关	line stereo matching
植被覆盖	vegetated surface
植被覆盖变化	vegetation cover change
植被光谱	vegetation spectrum
植被可视化	vegetation visualization
植被指数	vegetation index
植物冠层模型	canopy model
指北针	north arrow
指标差改正	correction for index error
指令跟踪站	command tracking station(CTS)
指派交叉	assignment crossover
指数	exponential
指数函数	exponential function
制动	clamp
制图分级	cartographic hierarchy

制图符号　cartographic symbol
制图简化　cartographic simplification
制图夸大　cartographic exaggeration
制图选取　cartographic selection
制图仪　cartograph
制图员　cartographer
制图专家系统　cartographic expert system
制图资料　source material, cartographic document
制图综合　cartographic generalization
质量感　qualitative perception
质量控制与评估　quality control and validation
质量评价　quality evaluation
质心　centroid
秩　rank
秩亏平差　adjustment with rank deficiency, rank defect adjustment
智能传感器网格　smart sensor web
智能交通系统　intelligent transportation system (ITS)
智能介质卡　smart media
智能可视化　intelligent visualization
智能摄影测量　intelligent photogrammetry
智能型全站仪,测量机器人　robotic (motorized) total station
智能主体　intelligent agent
置零,归零　adjust to zero
置信度　confidence, confidence flag

Z

置信极限	confidence limit
置信区间	confidence interval
置信水平	confidence level
置信椭圆	confidence ellipse
中比例尺	medium scale
中程定位系统	medium-range positioning system
中地地理信息系统软件	MapGIS
中地球轨道	medium Earth orbit (MEO)
中轨道卫星	mean earth orbit satellite
中国 SuperSoft 公司的卫星遥感测图处理系统	VirtuoZo
中国测绘学会	Chinse Society for Geodesy Photogrammetry and Cartography (CSGPC)
中国地理空间信息协调委员会	China Interagency Coordinating Committee on Geo-spatial Data
中国地理信息系统协会	China Association for Geographic Information System(CAGIS)
中国地理学会	Geographical Society of China (GSC)
中国国家基础地理信息中心	national geomatics center of China (NGCC)
中国资源卫星应用中心	China Center for Resources Satellite Data and Application (CRESDA)
中海达 GPS 数据处理软件	HDS2000/HDS2003

中丝	central cross-hair
中天	meridian transit, transit
中天法	transit method
中误差	mean square error(MSE)
中线测量	center line survey, location of route
中线	central line
中心城市	central city, key city
中心式快门	between-the-lens shutter, lens shutter
中心投影	central projection
中性色调,灰色调	middle tone
中央子午线	central meridian
中值定理	mean value theorem
中值滤波	median filter
终边	terminal side
终点	terminal point
终端框	terminal box
终端速度	terminal velocity
钟差	clock bias
钟偏	clock offset
钟速	clock rate
钟形曲线	bell-shaped curve
重采样	resampling
重氮复印	diazo copying
重点工程,大型项目	major project
重点实验室	key laboratory
重叠度,交叠	overlap
重叠影像法	overlapping image method

重复观测	repeated observation
重复试验	repeated trials
重合	coincide
重建,恢复,偿还	restitute
重力	gravity
重力[点]网	gravity network
重力测量	gravity measurement
重力测量卫星	gravisat
重力测量学	gravimetry
重力场	gravity field
重力场恢复	gravity recovery
重力垂线偏差	gravimetric deflection of the vertical
重力垂直梯度	vertical gradient of gravity
重力大地水准面	gravimetric geoid
重力点	gravimetric point, gravity station
重力方向	direction of gravity
重力固体潮观测	gravity observation of Earth tide
重力归算	gravity reduction
重力基线	gravimetric baseline, gravity datum
重力加速度	acceleration of gravity, gravity acceleration
重力矩	gravitational torque
重力控制点	gravimetric control point
重力偏差	gravimetric deflection
重力扰动向量	gravity disturbance vector, gravity equipotential surfaces
重力数据库	gravity database

重力水平梯度	horizontal gradient of gravity
重力水准测量	gravimetric leveling
重力梯度测量	gravity gradient measurement, gravity gradiometry
重力梯度仪	gravity gradiometer, gradiometer
重力卫星	CHAMP, GOCE, GRACE
重力位	gravity potential
重力向量	gravity vector
重力仪	gravimeter
重力仪的气压影响	atmospheric effect of gravimeter
重力仪的温度影响	temperature effect of gravimeter
重力异常	gravity anomaly
重力异常向量	gravity anomaly vector
重心	centre of gravity
重新照准	recollimation
重组织	reorganization
周长,周界	perimeter
周计数	time-of-week(TOW)
周年磁变	annual magnetic change
周年等磁变线	equivalent line of annual magnetic variation
周年视差	annual parallax
周年岁差	annual precession
周期	period
周期函数	periodic function
周期摄动	periodic perturbation
周期图	periodogram
周期误差	periodic error

Z

周日视差 diurnal parallax
周日运动 diurnal motion
周跳 cycle slip
周跳探测与修复 cycle slip detection and repair
轴 axis
逐次逼近法 successive approximation
主被动相关性 correlation of active and passive
主成分分析，主分量分析 principal component analysis (PCA)
主成分像元分解 principal component pixels decomposition
主垂面 principal plane [of photograph]
主动轮廓模型 active contour model
主动声呐 active sonar
主动式传感器 active sensor
主动式定位系统 positive positioning system
主动式遥感 active remote sensing
主动视觉 active vision
主动微波遥感 active microwave remote sensing
主动微波遥感传感器 active microwave sensors
主对角线 main diagonal
主分量变换 principal component transformation
主光轴 optical axis
主合点 principal vanishing point
主核面 principal epipolar plane
主核线 principal epipolar line
主机航速 revolution per minute(PRM)
主检比对 main/check comparison

主角	principal angle
主距	principal distance
主控站	master control station
主罗经	master compass
主台	main station
主网	principal network
主要点测设	setting out principal points
主要功能	key function
主要街道	major street
主轴	main axis, principal axis
主轴线测设	setting-out of main axis
助航标志	aid to navigation
助曲线,辅助等高线	extra contour
注册工程师	licensed engineer
注册建筑师	registered architect
注记	annotation
注入站	up-link station
柱面坐标	cylindrical coordinate
专家系统	expert system
专家知识	expert knowledge
专题测图仪	thematic mapper
专题层	thematic overlap
专题地图集	thematic atlas
专题地图学	thematic cartography
专题海图	thematic chart
专题图	thematic map
专题制图	thematic mapping
专题属性	thematic attribute

专用道路	special road
专用地图	special use map
专用海图	special chart
专属经济区	exclusive economic zone
转点	turning point
转动惯量	momentum of inertial
转发器,脉冲转发机	transponder
转换方程	transformation equation
转换器,交换器,换能器,变频管,变频器,转换反应堆	converter
转接器,拾音器,接合器	adaptor
转弯半径	turning radius
转折线	breakline
转置矩阵	transpose of matrix
桩	peg,stake,pile
桩,标桩	stake
桩基础	pile foundation
准确度	accuracy
准确度测试	accuracy testing
准实时纠正	real time rectification
准则	criterion
姿态	attitude
姿态参数	attitude parameter
姿态测量遥感器	attitude-measuring sensor(AMS)
姿态偏离	attitude deviation
姿态确定和轨道测量电子设备	attitude and orbit determination avionics(AODA)

资料替补	data substitution
资源容量	capacity
资源需求量	resource demand
资源与环境遥感	remote sensing for natural resources and environment
资源支持体系	resource support system
子(序)列	sub-sequence
子集	subset
子空间	subspace
子午方向	meridian direction
子午高度角(天体中天的高度角)	meridional height
子午面	meridian plane, meridian section
子午圈	meridian
子午圈曲率半径	meridian radius of curvature
子午线弧	meridian arc
子午线曲率	meridian curvature
子午线曲率改正	correction for meridian curvature
子午线收敛角	convergence of meridine, meridian convergence
子午线收敛角,坐标纵线偏角	meridian convergence, grid convergence
子像素,亚像素	subpixel
子像素匹配	subpixel matching
字节	byte
自动安平水准仪	automatic level, compensator level
自动报警控制	automatic alarm control
自动编组	grouping

Z

自动标定	auto-calibration
自动补偿器	automatic compensator
自动化地图制图	automatic cartography
自动绘图	automatic plotting
自动绘图机	automatic plotter
自动绘图系统	automated drafting system
自动剪切	automatic clipping
自动检测	automatic detection
自动卷片	automatic film advance
自动空中三角测量	automatic aerial triangulation
自动目标识别	automatic target recognition (ATR)
自动配准	automatic registration
自动匹配	automatic match
自动识别	automatic identification, automatic recognition
自动数据传输	automatic data transmission
自动搜索	automatic search
自动探测	automated detection
自动镶嵌	auto-mosaicking
自动制图	automated mapping
自动制图软件	automated cartography software
自动制图-设施管理系统	automated mapping/facilities management system(AM/FM)
自动制图系统	automated cartographic system
自动制图综合	automated cartographic generalization
自动坐标展点仪	automatic coordinate plotter

Z

自反关系	reflexive relation
自回归参数	autoregressive parameters
自回归滑动平均模型	autoregressive-moving average model(ARMA model)
自回归模型	autoregressive model(AR model)
自检校	self-calibration
自检校区域网平差	selfcalibrating block adjustment
自来水	city water
自然保护区	nature reserve
自然表面	physical surface
自然地图	physical map
自然对数	natural logarithm
自然光	natural light
自然环境	natural environment, physical enviroment
自然排水	natural drainage
自然数	natural number
自然通风	natural draft
自然灾害	national hazards
自适应	adaptation
自适应共振理论	adaptive resonance theory(ART)
自适应量化	adaptive quantization
自适应滤波	adaptive filter
自适应视差估计	adaptive disparity estimation
自适应算术编码	adaptive arithmetic coding
自适应梯度	self-adaptive gradient
自适应最小二乘相关	adaptive least squares correlation
自相关	autocorrelation

自相关函数	autocorrelation function
自相似	self-similarity
自协方差	autocovariance
自协方差函数	autocovariance function
自由导线,支导线	free traverse
自由度	degrees of freedom
自由设站法	free station
自由下坠	free fall
自由向量	free vector
自重	own weight
自转轴平极	mean pole of rotational axis
自准直目镜	autocollimating eyepiece
自准直主点	principal point of autocollimation (PPA)
自组织	self-organization
自组织特征映射	self-organizing feature map
自组织网络	self organizing network
渍害	waterlog damage
宗地	parcel
宗地图	parcel map
综合地图	comprehensive map
综合地图集	comprehensive atlas
综合规划	comprehensive planning
综合监测方法	integrated monitoring method
综合孔径	synthetic aperture
综合信息场	Conformity_congregate Information Field(CCIF)
总长度	overall length

总尺寸	overall dimension
总电子含量	total electron content(TEC)
总方量	aggregate volume
总高度	overall height
总建筑占地面积	gross building area
总居住区面积	gross residential area
总跨度	total span
总面积	gross area
总平面图	general loayout plan
总平面图，总体规划	comprehensive planning, master plan
总岁差	general precession
总体	population
总体规划	general planning,overall planning
总体精度估值	overall (global) precision estimates
总体精度准则	overall (global) precision criterion
总体平均(值)	population mean
总图	general map
纵断面	profile
纵断面测量	profile survey
纵断面水准测量	profile leveling
纵断面图	longitudinal profile,profile diagram
纵横比例校正	correction for scale distortion
纵截面	longitudinal section
纵坡	longitudinal gradient
纵倾	trim
纵视图	longitudinal view

Z

纵摇	pitch
纵坐标	ordinate
纵坐标轴,Y 轴	ordinate axis
走行式声学海流计	acoustic hull mounted current meter
走走停停定位法	stop-and-go positioning
阻碍强度	impedance
阻浮器	depressor
阻尼简谐运动	simple damped harmonic motion
阻尼振动	damped oscillation
组分温度	component temperature
组分温度反演	inversion of component temperature
组合	combination
组合导航	integrated navigation
组合导航系统	integrated navigation system
组合地图	homeotheric map
组合定位	integrated positioning
组合分类	combination classification
组合航法	composite sailing
组件式地理信息系统	component GIS
组界	class boundary
组区间	class interval
钻孔	borehole, drilling
钻孔爆破法	drill & blast method
钻孔机	drilling machine
钻孔位置测量	borehole position survey
最大荷载	ultimate load
最大后验概率	maximum a-posteriori probability

最大静摩擦	limiting friction
最大绝对误差	maximum absolute error
最大谱图像配准	max-spectrum image registration
最大熵	maximum entropy
最大似然分类	maximum likelihood classification
最大似然估计	maximum likehood estimation
最低低潮面	lowest low water
最短路径	shortest route
最或然值	most probable value(MPV)
最佳基	best basis
最佳拟合	line of best-fit
最近邻迭代	iterative closest point algorithm (ICP)
最浅水深	shallowest sounding
最少成本路径	least-cost path
最小二乘法	least squares method(LSM)
最小二乘估计	least squares estimation
最小二乘理论	theory of least squares
最小二乘模板匹配	least squares template matching
最小二乘配置法,最小二乘拟合推估法	least squares collocation
最小二乘匹配	least squares matching
最小二乘谱	least squares spectrum
最小二乘谱分析	least squares spectral analysis
最小二乘相关	least squares correlation
最小二乘影像匹配	least squares image matching
最小方差无偏估计	minimum variance unbiased estimation
最小距离分类	minimum distance classification

最小限定矩形	minimum bounding rectangle
最小约束	minimum constraints
最小约束平差	minimum constraint adjustment
最小制图单元	minimum mapping unit
最优波段组合	optimum wave-band combination
最优超平面	optimal hyperplane
最优解	optimal solution
最优设计	optimum design
最优阈值化	optimal threshold
左侧	left femur
左方极限	left-hand limit
左手边	left-hand-side(L. H. S)
左手坐标系	left-handed coordinate system
左右视差	horizontal parallax, x-parallax
坐标	coordinate
坐标北方向	grid north
坐标变换	coordinate conversion
坐标差	coordinate difference
坐标地籍	coordinate cadastre
坐标法	coordinate method
坐标方位角	grid bearing
坐标方位角，格网方位角	grid azimuth
坐标量测仪，比长仪，检定器	comparator
坐标平差	coordinate adjustment
坐标系	coordinate system
坐标协因数矩阵	cofactor matrix of coordinates

Z

坐标仪,直角坐标展点仪,绘图仪	coordinatograph
坐标原点	grid origin,origin of coordinate system
坐标增量	coordinate increment
坐标增量闭合差	closing error in coordinate increment
坐标中误差	mean square error of coordinate
坐标轴	coordinate axis
坐标轴,基准轴	reference axis
坐标转换	coordinate transformation

参考文献

1. 测绘学名词审定委员会. 测绘学名词. 北京:科学出版社,2002.

2. 《英汉测绘词汇》编辑组编. 英汉测绘词汇. 北京:测绘出版社,1982.

3. 航海科学名词审定委员会. 航海科技名词. 北京:科学出版社,1996.

4. 地理信息系统名词审定委员会. 地理信息系统名词. 北京:科学出版社,2002.

5. 孔令户等. 海洋测绘词典. 北京: 测绘出版社,1999.

6. 尹晖等. 测绘工程专业英语. 武汉:武汉大学出版社,2005.

7. 张祖勋,张剑清. 数字摄影测量学. 武汉:武汉大学出版社,1997.

8. 孙家抦. 遥感原理与应用. 武汉:武汉大学出版社,2003.

9. 关泽群,刘继林. 遥感图像解译. 武汉:武汉大学出版社,2007.

10. GB/T 17694-1999,地理信息技术基本术语.

11. Barry F. Kavanagh . Geomatics. Pearson

Education Inc,2003.

12. W. Schofield . Engineering Surveying: theory and examination problem for students. 5th Ed. Butterworth-Heinemann,2001.

13. Paul R. Wolf and Charles D. Ghilani . Adjustment Computations: Statistics and Least Squares in Surveying and GIS. New York: John Wiley and Sons, Inc,1997.

14. Barry F. Kavanagh and S. J. Glenn Bird . Surveying: principles and applications. 4th Ed. Prentice-hall Inc,1996.

15. Hofmann-Wellenhof B, Lichtenegger H, Collins J. Global positioning system: theory and practice. New York: Springer-Verlag, 2000.

16. Hofmann-Wellenhof B, Legat K, Wieser M. Navigation-principles of positioning and guidance. Springer, Wien New York, 2003.

17. Leick, A.. GPS satellite surveying, Second Edition. New York: John Wiley and Sons, p. 584. ISBN: 0471306266, 1995.